The Barefoot Beekeeper

A simple, sustainable approach to
small-scale beekeeping
using top bar hives

by

P J Chandler

www.biobees.com

The Barefoot Beekeeper

Read not to contradict and confute,
nor to believe and take for granted,
nor to find fault and discourse;
but to weigh and consider.
Francis Bacon

Third Edition
Spring 2009

ISBN 978-1-4092-7114-7

The Barefoot Beekeeper

Preface to the Third Edition

This book is for prospective, beginner and experienced beekeepers, who want a simple method of looking after a few hives for small-scale production of honey, beeswax, pollen and propolis – or just for the fun of having some bees around.

This does not pretend to be a complete work on beekeeping and it assumes that at least some knowledge of the subject has already been acquired from other sources and that you are prepared to read more widely. There are links to beekeeping bibliographies on my web site.

You will learn most from your own experiences, but these take time to acquire and you should take every opportunity to learn from others – especially others' mistakes – as this will save you time, money and frustration.

My aim is to demonstrate that beekeeping does not need to be difficult, time-consuming, expensive or needlessly complicated and that almost anyone – including people with disabilities and mobility problems – can learn about, practise and benefit from this fascinating and absorbing activity. Everything needed for 'barefoot beekeeping' can be made at home using hand tools and only basic skills.

It is written by an English beekeeper and while the principles are universal, local climate, flora, seasonal weather conditions and experience will dictate variations in your approach that should be followed more assiduously than anything written here.

Readers are encouraged to join our worldwide 'natural beekeeping' forum, where ideas and knowledge are freely shared.

www.biobees.com

ACKNOWLEDGEMENTS

My thanks for help and support in this project go to: my partner Lesley for her patience, understanding and unwavering encouragement; John Phipps, editor of The Beekeepers Quarterly, for his helpful critique of my draft and for several photographs; Marty Hardison for inspiration, helpful and very constructive criticism and photographs; Andria King, John Rippon, Norm and Gary for their suggestions and corrections; Brian Gant for helping me with my very first beehive and his expert advice over the years.

Since the first edition was published in early 2007, I was contacted by Dr David J. Heaf, of the Science Group of the Anthroposophical Society in Great Britain who, with his wife, has translated Abbé Warré's book, *Beekeeping For All* into English from the original French. This has added a new, vertical dimension to our thinking about the top bar hive and and I hope that cross-fertilization between the two systems will lead to greater understanding of truly sustainable beekeeping.

DEDICATION

This book is dedicated to the honeybee. All proceeds from the sale of this book will be used to further the development of sustainable beekeeping.

If you bought this book, thank you for your support. If you did not buy it, you are invited instead to make a donation to the project at www.biobees.com.

Contents

INTRODUCTION

Since the turn of the 21st century, I have kept bees in WBC hives, skeps, home-made framed hives and, since 2004, exclusively in top bar hives. I spent a full year working with the bees at Buckfast Abbey, where I was privileged to be able to read the late Brother Adam's collection of beekeeping books, study his bee breeding methods and work with what remained of his bees.

Some will say that this is far too little time in which to gather sufficient experience in the craft to make any worthwhile contribution to the ever-growing mountain of literature on the subject.

They are right, of course: I doubt that I will know enough about bees even in another ten or twenty years to feel truly confident about my pronouncements, but such is the woeful state of bees and beekeeping in the early years of this century that I offer these thoughts to those who care to listen, in the hope that we can do enough, quickly enough, to save the bees from what appears to be terminal decline[1].

Why do I, with not so much as a first science degree, believe that I have the answers to the ills of bees?

Firstly, I do not claim to have all the answers. Few - if any - of the ideas presented here are unique to me, nor do I claim any particular originality for the beekeeping methods I describe. The particular top bar hive design illustrated here is my own, but it is really only a development of traditional African (and before that, Greek and Egyptian) top bar hives and differs from those of other top bar beekeepers only in a few points that I consider innovative and important, but others may not.

There is nothing really new in beekeeping – only old ideas recycled in new clothes.

Secondly, I invite you to consider what I and others have to say before drawing your own conclusions. I believe, having seen the evidence with my own eyes, that current 'standard' beekeeping methods - together with our toxic, chemical-based agricultural system - are responsible for most of the problems suffered by our bees. I also believe, having performed some crude experiments and having spoken at some length

---

1 Rudolf Steiner warned in 1923 that beekeeping would become unworkable within 50 to 80 years. Abbé Warré recognized the decline too. Johann Thür, Bee-keeper (Wien, Gerasdorf, Kapellerfeld) in his book *Bienenzucht. Naturgerecht einfach und erfolgsicher.* (2nd edition, 1946) described the decline – i.e. bee diseases – and blamed, above all, the use of frames. Thür argued that warmth is a hive's most valuable asset after food and that the law of *Nestduftwärmebindung* – retaining the nest warmth and atmosphere (humidity, pheromones, and possible volatile compounds connected with nest hygiene) – should not be violated. It is less violated in a long format TBH than in a framed hive. *(Comment contributed by Dr David J. Heaf, Newsletter & "Archetype" Editor, Science Group of the Anthroposophical Society in Great Britain.)*

with others working along similar lines, that the way forward is to work more closely *with* the bees, developing a relationship based on mutual benefit and co-operation rather than simple exploitation.

The author with one of his hives at Riverford Farm in Devon

And if all that sounds too 'new age' for some, let me also add that I am, above all, a practical man looking for real-world solutions.

My primary aim in writing about 'barefoot beekeeping' is to challenge the status quo and to stimulate both actual and potential beekeepers to think for themselves and to ask more questions.

Framed hives have been the accepted standard for more than a century: only a handful of beekeepers have challenged their ubiquity, yet the beekeeping 'establishment' continues largely to ignore alternatives, despite the obvious and manifold drawbacks of the system they promote.

Newcomers to beekeeping are the most likely to approach the subject with an open mind, and I hope they will find appealing the ideas that are presented here.

My secondary aim is to describe a sustainable beekeeping system based on simple principles.

THE PRINCIPLES OF BAREFOOT BEEKEEPING

There are too many 'books of rules' in the world already, and I have no intention of writing another one. Instead, I wish to propose three basic principles, which form the basis of the 'barefoot' approach to beekeeping:

1. **Interference in the natural lives of the bees is kept to a minimum.**

2. **Nothing is put into the hive that is known to be, or likely to be harmful either to the bees, to us or to the wider environment and nothing is taken out that the bees cannot afford to lose.**

3. **The bees know what they are doing: our job is to listen to them and provide the optimum conditions for their well-being.**

'Barefoot beekeeping' is for both urban and rural dwellers who want to keep bees on a modest scale, producing honey and beeswax (and, perhaps, propolis and pollen) for their own use and for friends and neighbours. This is not intended to be a blueprint for large-scale, commercial bee farming, which I believe to be very much part of the problem.

All equipment is designed to be built using sustainably grown, low-cost materials by people with only moderate manual skills: if you can do a decent job of putting up shelves, then you can probably make a serviceable beehive. Bees are very forgiving of imperfect joinery.

Above all, 'barefoot beekeeping' is for people who love bees and understand and appreciate their vital role in the pollination of a huge range of both wild and cultivated plants.

Philip Chandler

(I should mention - for the benefit of readers with a literal turn of mind – that the term 'barefoot' is merely a metaphor, intended to convey an attitude of simplicity in one's approach to the subject. I do not suggest that you do your beekeeping sans footwear.)

WHY DO YOU WANT TO KEEP BEES?

If your main aim is to obtain the absolute maximum amount of honey from your hives, regardless of all other considerations, then you are reading the wrong book. Not that this style of beekeeping cannot produce decent amounts of honey – it certainly can – but the emphasis here is on *sustainability* and *keeping healthy bees* rather than setting records for honey crops, which inevitably has a cost to the welfare of the bees.

The essence of sustainability is to work well within the limits of a natural system: pushing any living thing beyond its natural capacity can only lead to trouble.

Let me lay my cards on the table right away: I believe that beekeeping should be a small-scale, 'cottage industry', or part-time occupation or hobby and should be carried out in the spirit of respect and appreciation for the bees and the part they play in our agriculture and in nature. I disapprove of large-scale, commercial beekeeping because it inevitably leads to a 'factory farming' mentality in the way bees are treated, handled and robbed and a lack of consideration of its effects on biodiversity.

Bees evolved to live in colonies distributed across the land according to the availability of food and shelter. Forcing 30, 50, 100 or more colonies to share the territory that, perhaps half a dozen would naturally occupy is bound to lead to concentrations of diseases and parasites that could not otherwise occur and that can only be dealt with by means of chemical or mechanical interventions, which, I and many others believe, weaken the bees' natural defences.

Bees love to feed on a multiplicity of flowers, as can be easily demonstrated by the variety of different pollens they will collect if sited in a wild place with diverse flora. Transporting them to a position where there is only a single crop of, say, oilseed rape within reach prevents them from exercising their desire for diversity and causes an unnatural concentration within the hive of a single pollen, which is most likely lacking in some of the elements they require for full health, and may also contain small amounts of toxic pesticides[2], which, when gathered into the hive may reach lethal concentrations. Yet migratory beekeeping is practised in just this way on an industrial scale in some countries, especially the USA.

---

2 Of particular concern at the moment are the neo-nicotinoids, which have been shown to be lethal to bees in microscopic quantities, perhaps as low as 5 parts per billion. This is equivalent to one teaspoon of pesticide in one thousand metric tonnes of nectar.

From a conservation point of view, unnaturally large concentrations of honeybees can also threaten the existence of other important and, in places, endangered pollinating insects, such as bumble bees and the many other species that benefit both wild and cultivated plants.

Sustainable beekeeping is small-scale by definition. It is 'backyard beekeeping' by people who want to have a few hives at the bottom of their garden, on their roof (there are a surprising number of roof-top beekeepers in our cities) or in their own or a neighbour's field or orchard.

Probably you want to produce modest quantities of honey for your family and friends, with, maybe, a surplus to sell at the gate or in the local market. You will have by-products; most obviously beeswax, which you can make into useful stuff like candles, skin creams, wood polish and leather treatments, so beekeeping could become the core of a profitable sideline.

And you are interested in bees for their own sake, I hope. If not yet, I have no doubt that you will be once you have looked after a few hives for a season or two.

You may have been to an open day hosted by your local beekeeping association, or read a book or two, or perhaps you have taken the plunge already and bought a second-hand hive and captured a swarm or obtained a 'nuc'[3]. You may have browsed through the catalogues of beekeeping suppliers, wondering at the enormous number of specialized gadgets and pieces of equipment you seem to need and wondering where you would put it all and how you would pay for it.

In this case, you will be truly thankful to know that my mission throughout this book is to show you that, (a) beekeeping does not have to be as complicated as some would make it out to be and (b) you need *none* of the stuff in those glossy beekeepers' supplies catalogues in order to keep healthy, happy and productive bees.

None of it at all.

You will recall that the sub-title of this book is '*A simple, sustainable approach to small-scale beekeeping*' and that is what I have in mind throughout and I would like you to keep in mind: *simple, sustainable, small-scale.*

The system I will describe here is about as simple as beekeeping can get, while maintaining provision for occasional inspections, comfortable over-wintering and non-destructive harvesting. Everything you need is in one box – the beehive – which you can make yourself if you follow my

---

3 A nucleus – or nuc - hive usually comprises 3-5 frames of bees in a suitably sized box, with a laying queen.

instructions. TBH plans are available elsewhere, but naturally I believe mine to have certain advantages, which I trust will become clear as you read on.

You can – indeed you should - buy or make yourself a veil. If you are nervous, you could even get a beekeeper's suit or a smock, but any light-coloured shirt will do as well. A hive tool can be handy, but a strong, sharp, flat-bladed knife will also work.

Some of the things you will *not* need include:

- frames
- foundation wax
- supers
- centrifugal extractor
- bottling equipment
- de-capping knife and tray
- bee escapes
- mouse guards
- queen excluders
- fancy feeders
- space suits
- bee blower

And you probably won't need gloves or a smoker, but if you already use them, or are nervous of bees, then by all means use them if they help you to feel more confident.

What you will need is a hive – probably two or three or more in time – and I will show you how to build them cheaply and easily, using only hand tools if you prefer, with only rudimentary woodworking skills. You will find fully-illustrated instructions in my downloadable ebook called, 'How To Build a Top Bar Hive', obtainable free in several formats from my web site: www.biobees.com.

Bees are fascinating creatures and among the many beekeepers I know or have talked to – even commercial men - I can't think of any who keep them solely for the income they generate.

A Greek-style top bar basket hive

So be warned: if you start keeping bees and develop a real interest in them, it will be with you for life. And I doubt very much that you will regret it for a moment.

But before we get into the practical stuff, a little background.

A BRIEF HISTORY OF BEEKEEPING IN BRITAIN AND THE USA

Honeybees, in something like their present form, have been around for at least 40 million years[4]. In that time they have evolved into one of the most successful and highly organised social creatures on earth.

We humans have been around for only about three million years and probably only in the last few thousand have we developed a relationship with bees, largely consisting of us finding new and more creative ways of robbing and exploiting them.

Primitive hives made from logs, baskets and pots of various kinds were – and in some places still are – used to provide homes for bees, while offering more-or-less convenient means by which their honey could be removed as required. In the UK and much of Europe, straw skeps of many and varied designs were the standard hive for centuries and were in common use right up until the middle of the twentieth century. I know of a group of beekeepers in Germany who still use skeps on the heather and there are still some dedicated skeppists dotted about Britain.

Straw skep

For an unknown period – perhaps 1000 years or more – beekeeping in Britain was carried out mainly by monks and peasant farmers, usually in straw or rush skeps. Swarming was the principle method of maintaining a stock of colonies, prime swarms being captured and housed as they emerged or as soon as they could be caught. A certain number of colonies were killed off at the end of each season in order to extract their honey and wax comb, as no means was then available for non-destructive harvesting. There were plenty of wild colonies around, which provided a reservoir of new blood, strengthened by the process of natural selection. The best managed colonies were overwintered and swarms emerging from them in the following season ensured a plentiful supply of bees for everyone.

---

4 I quoted a much earlier date in previous editions: this now seems to be the generally agreed figure.

MOVABLE FRAMES: THE HOLY GRAIL

The advent of modern beehives and their associated technology during the latter half of the nineteenth century made the processes of bee management and honey extraction easier and more efficient and laid the foundations for industrial-scale, commercial beekeeping as we see it today. The hobby beekeeper was also able to take advantage of this new technology, resulting in many enthusiastic amateurs keeping a couple of hives at the bottom of their gardens. A new breed of beekeeper emerged among the clergy and middle classes, driven by the scientific and industrial impulse of the Victorian era, who sought new ways to control this fascinating wild creature and bend her behaviour to the needs and desires of man.

Langstroth's original hive

The key invention that made all this possible was the self-spacing, 'movable frame', introduced by the Rev. L. L. Langstroth around 1850 in the USA, although his work owed a great debt to Jan Dzierzon[5]. Wooden frames, arranged side by side across the width of a rectangular box, spaced apart according to the recent discovery of 'bee space'[6], meant that bees could conveniently be manipulated and 'managed' as never before, according to the various theories and whims of beekeepers.

Because – according to some accounts - Langstroth had chosen a discarded wine case that just happened to by lying around in his workshop on which to base his 'standard' (which remains to this day as the standard dimensions for the Langstroth hive), the shape of his frame was that of a rectangle approximately twice as wide as it was deep – utterly unlike the tall, catenary curves of the comb that bees like to build when left to their own devices. Nevertheless, bees are versatile and flexible and they adapted themselves as best they could to the new shape.

In case you think that I have any illusions that beekeeping before the middle of the 19th century was some kind of perfect idyll, here is what

---

5 Jan Dzierzon (16 January 1811 – 26 October 1906) was a Polish apiarist and Roman Catholic priest who was best known for his discovery of parthenogenesis among bees, and for designing the first successful movable-frame beehive.

6 The working space that bees leave between combs: between 1/4" and 3/8"

Langstroth himself had to say about it:

> *The present condition of practical beekeeping in this country [the USA] is known to be deplorably low. From the great mass of agriculturists, and others favourably situated for obtaining honey, it receives not the slightest attention. Notwithstanding the large number of patent hives which have been introduced, the ravages of the bee-moth have increased, and success is becoming more and more precarious. Multitudes have abandoned the pursuit in disgust, while many of the most experienced are fast settling down into the conviction that all the so-called "Improved Hives" are delusions, and that they must return to the simple box or hollow log, and "take up" their bees with sulphur, in the old-fashioned way.*[7]

In Britain, where it seems that nearly every Victorian beekeeper considered himself an inventor, there was less standardization of dimensions and any number of variations arose on the theme of movable frames in a box. Notable among these was the WBC hive, invented by one William Broughton Carr about 1890 and still around today in a limited way. Having featured in innumerable children's books it is the shape the general public most readily associate with beehives and is still often (and inexplicably) recommended to beginners, despite its prodigious use of timber and considerable 'nuisance factor' due to the extra lifting, maintenance and storage required.

The WBC and its innumerable variations have largely given way among hobbyists to the British National Hive, which, though dull and functional in appearance compared to the WBC, is more restrained in its use of timber and therefore cheaper, lighter and more practical in many ways. There is a deep frame variant of the National, which has its adherents, while the capacious Commercial hive is favoured by larger-scale beekeepers. The only other notable 'modern' British hive is that built to the specifications of the late Brother Adam at Buckfast Abbey in Devon. Bro. Adam's 'Modified Dadant' hive is a monster, some 20 inches square by 12” deep and containing up to 12 frames, each having roughly twice the brood area of a National (14" x 8 ½") frame. Moving these hernia-inducing boxes and their accompanying supers requires considerable strength (Bro. Adam had a free labour force of monks at his disposal) and no hobbyist need give them a second look, save from curiosity. The Langstroth hive is little used in

The WBC Hive

7 Rev. L L Langstroth, The Hive and the Honey Bee, 1853

Britain, although ubiquitous in the USA, Canada and many other countries.

Both in Britain and in North America, along with the rest of the developed world, movable frames fitted with wax foundation to a standard pattern became the unquestioned orthodoxy of beekeeping. New beekeepers acquired equipment and knowledge from old beekeepers in the same way that apprentices learned their trades from master craftsmen and thus perpetuated the status quo right up to the present day.

The next important invention that handed yet more control to the beekeeper was that of pre-fabricated wax 'foundation'. It was considered that bees spent too much of their time and energy (and, therefore, honey) on building wax comb and, if they could be 'helped along' by the provision of thin sheets of wax, impressed with a suitable hexagonal pattern, pre-fitted to the wooden frames, then that could only be a Good Thing.

Because the embossed pattern was designed to emulate the beginnings of worker cells, bees were thus forced to fill their homes with worker comb and were discouraged from making 'useless' drone cells. Foundation was made according to measurements made by A I Root around 1884 and, largely due to Root's ubiquity in the US beekeeping supplies market, seems to have been milled more or less to these dimensions to this day.

Bees will take any opportunity to build drone cells in odd corners and often they will build a whole comb of them against one of the internal walls, despite the beekeeper's efforts to thwart them. The general practice among beekeepers is to prevent their bees from raising 'too many' drones by culling drone brood: a maximum of 5% seems to be the accepted figure. The thinking is that drones, being unproductive and having no obvious work to do aside from mating, must therefore be supernumerary and dispensable. They also consume honey, of course – a lot according to some and hardly any according to others – but that is often given as a reason to cull them. Left to their own devices, bees will ensure that, in the queen mating season, they have up to 20% of their number as fertile males (drones). This discrepancy may, I suggest, be a major factor in the recent reports of many queens failing to mate or being poor layers and has almost certainly accelerated the spread of the feral 'Africanized' bees[8] in the USA, which are not subject to the whims of beekeepers and can flood an area with their own drones with little competition from hived bees.

I think it is more than likely that drones do in fact have other functions

---

8 An unfortunate cross between Apis mellifera scutellata and A. m. ligustica

within the hive. In particular, I think they have a role to play in maintaining the correct hive temperature for the brood. Remember that, in temperate climes, the inside of the hive – especially in the main brood area - is *always* warmer than the outside world: around 94°F (34°C), a temperature they maintain throughout the year with little variation[9]. *This means that opening a hive at any time of year will cause the bees a deal of extra work in returning their environment to its correct temperature* – a fact that receives barely a mention in any beekeeping book I have read, other than Abbé Warré's *Beekeeping For All*[10]. In hotter countries, opening the hive gives bees the opposite problem: how to cool it back down to their working temperature. This is a powerful argument for 'natural' or 'leave well alone' beekeeping in a hive designed to be managed in this way and an equally powerful argument against opening any hive unnecessarily - even in summer - and especially a hive that opens at the top exposing bees all at once, as is the case with all framed hives.

Conventional, framed hives create a lot of extra woodwork and a storage problem.

9 W. E. Dunham, Department of Zoology and Entomology, Ohio State University, Columbus, 1926.
10 The 12th (1948) edition translated into English by Dr. David Heaf (see warre.biobees.com)

In the 1940s, Johann Thür, a German beekeeper who favoured vertical top bar hives in the style of Abbé Warré, described in *Bienenzucht* his concept of *Nestduftwärmebindung.*[11] This introduces the notion of a combination of heat and scent that provides a beehive with its unique, nurturing and disease-resistant 'nest atmosphere', which should not be disturbed. In his view, it is incumbent upon us as beekeepers to respect the bees' need to maintain this 'nest atmosphere' and to design hives and management protocols that disrupt it as little as possible.

It is entirely possible that this combination of heat and scent may be important for the suppression of the *Varroa* mite, which, I have read, cannot reproduce above about 92° F (33° C). A recent study showed that undisturbed, feral colonies seemed better able to co-exist with *Varroa* mites than those managed in a conventional way.[12]

PESTS AND DISEASES

Apis mellifera probably evolved in Africa and spread across Europe and Asia over millions of years, adapting to a wide variety of climates. Each local variety has particular characteristics - some visible, some behavioural - that distinguish it from others.

Up until about 100 years ago, the ubiquitous native British bee was *Apis mellifera mellifera* – having a dark brown colour, shading into black. This was a hardy bee, that had evolved and adapted itself to suit the unpredictable British climate, capable of bringing in a good winter stock of honey even in a poor summer. It had a reputation for being quite defensive, which meant that it was less popular among beekeepers who used 'modern' movable-frame hives than among the old skeppists, who rarely handled their bees from the beginning of the season until the honey harvest.

During the late nineteenth and early twentieth century, honeybee colonies began to suffer on an unprecedented scale from a range of diseases and parasites that had previously been rare, localized or relatively mild in their effects.

By 1920, according to some sources, the native British black bee had been virtually wiped out by so-called 'Isle of Wight disease'[13], to which it

---

11 from *Bienenzucht. Naturgerecht einfach und erfolgsicher* by Johann Thür, Imker (Wien, Gerasdorf, Kapellerfeld, 2 ed., 1946) Translated by David Heaf

12 Seeley, T. D. (2007). Honey bees of the Arnot Forest: a population of feral colonies persisting with *Varroa destructor* in the north-eastern United States. Apidologie, 38: 19–29.

13 Now believed to have been a massive outbreak of Acarapis woodii, a tracheal mite, in which case

had no natural resistance. This turns out to have been an epidemic of an internal mite called *Acarapis woodii*, that was first found on the Isle of Wight in 1904 - presumably by someone using a microscope. Where they came from, nobody seems to know, but it could so easily have been the first example of a major varietal loss of a native British species due to nascent globalization: the real beginnings of the internationalization of the bee industry.

Replacement black bees were brought in from France, Germany and Holland, along with yellow-striped bees from Italy to re-stock the empty hives, but crosses between the black and yellow races were (and still are) overly defensive and difficult to manage. While they were much less susceptible to 'Isle of Wight' disease, the mild-mannered Italians, along with the other immigrants, were vulnerable to both American and European Foul Brood (AFB and EFB), the two most serious bee diseases. And they were (and still are) incurable robbers of other bee colonies. [14]

During the 20th century, various attempts were made to breed the 'perfect' bee, most notably by Brother Adam, a Benedictine monk of German origin, at Buckfast Abbey in Devon. He travelled widely to gather genetic material to incorporate into his famous 'Buckfast' strain. His goal was to produce a disease-resistant, good-tempered, manageable and productive bee with excellent over-wintering abilities and many beekeepers, particularly in Germany, Scandinavia and the USA will testify that, in its 'pure' form, the Buckfast Bee has all these qualities. However, if it is allowed to out-cross with random mongrels, the resulting progeny – while still inclined to be productive – are often very bad-tempered indeed[15].

For the 'pure strain' breeder, maintaining those desirable traits from generation to generation by a careful program of breeding is vital and – together with heterosis[16] - was the secret of Brother Adam's success and worldwide fame.

Despite the brave efforts of Bro. Adam and other breeders, bees continued to die in significant numbers from foul brood, acarine and *Nosema apis* (a microsporidian: a unicellular parasite) and new pests began to appear, most notably a parasitic mite, originally labelled *Varroa jacobsonii,* later changed to the more ominous-sounding *Varroa destructor.*

'tolerance', rather than 'resistance' would be a better term. There is evidence that A.m.m. survived and is still widespread throughout Britain and Ireland.

14 This was first noted – as far as I can establish – by A. Gilman in his 'Practical Bee-Breeding' 1928

15 This also seems to be the case with other 'pure-bred' strains.

16 Also known as 'hybrid vigour' – the tendency of hybrids to out-perform their parents.

Not yet in Britain (as of 2008), but nevertheless posing a longer-term threat, is the highly destructive Small Hive Beetle *Aethina tumida*, and another genus of parasitic mites similar in habit to *Varroa*, called *Tropilaelaps*. These mites are carriers of several viruses potentially lethal to bees and the mites themselves weaken their hosts by feeding on haemolymph, the bees' 'blood'. Kashmir virus, probably carried by mites, has recently been detected (2005) in two colonies in the north of England. Apparently, we can also expect an invasion of giant Japanese hornets from France. There has been much talk in 2007-8 of Israeli Acute Paralysis Virus (IAPV) gaining a hold.

Honeybees, on which we depend for the pollination of so many of our food crops, are now in trouble as never before and much of the blame for this potentially disastrous state of affairs must be placed at the door of negligent, commercial beekeepers. The inter-continental migration of pests and diseases has widely been blamed on climate change, but in fact the spread of the *Varroa* mite from its native Asia and its original host species, the Asian bee *Apis cerana,* can be directly linked to the commercial bee trade. Current (2008) reports of the widespread occurrence of *Nosema ceranae* demonstrate that such lessons are not quickly learned.

> *Ectoparasitic mites of the genus Varroa are known from Asian honey bees, of which nine extant Apis species are recognised (Koeniger and Koeniger 2000). All life stages of Varroa mites feed exclusively on bee haemolymph after perforating the host's integument with their chelicerae (Smirnow 1979;Donze and Guerin1994). The so-called western honey bee, Apis mellifera, with 24 subspecies distributed over Europe, Africa and the Near East (Ruttner 1988), has been repeatedly infested with Varroa destructor during the last century. This occurred through contacts with the closely related Apis cerana as a consequence of the worldwide transport of bee colonies and apicultural projects in developing countries (Matheson 1993). Today Varroatosis is the main problem for beekeeping with A. mellifera colonies (De Jong 1997).* [17]

Varroa probably co-existed with *Apis cerana* for many thousands of years and in that time the two species reached an accommodation whereby the bees learned to keep the parasites down to a tolerable level without actually eradicating them. When, thanks to the activities of bee-keepers in their home area, *Varroa destructor* came across our honeybee, *Apis mellifera,* it found a new and vulnerable host, which had had no opportunity to evolve a defence mechanism. Honeybees began to die in their millions as the mites exploited their new hosts' susceptibility and spread across the globe with astonishing rapidity.

An effective treatment was found in the form of the synthetic, miticidal pyrethroid *fluvalinate* and to some extent the *Varroa* mite was brought

---

17 G. Kanbar Æ W. Engels, Ultrastructure and bacterial infection of wounds in honey bee (Apis mellifera) pupae punctured by *Varroa* mites, published online: 27 March 2003

under control. However, within a few years the mites evolved a resistance to fluvalinate, aided by some beekeepers who, through laziness or incompetence, applied a low-level dose over a period of months instead of a calculated dose over a few weeks. Regrettably, I have myself seen, in a commercial apiary, Apistan (fluvalinate) strips that had been left in hives for as much as nine months.

Nobody knows for sure how *Varroa* arrived in Britain, but it was first detected by an amateur beekeeper at Torquay in Devon, in 1992, which could indicate importation on a Channel Islands ferry or a fishing boat, although the fact of its discovery there may simply mean that it had not previously been noticed elsewhere. It has since spread throughout the British Isles and by the summer of 2005, mites with resistance to pyrethroids were distributed across south west England, south Wales and elsewhere, with patches as far north as Durham. We can now, I think, presume that most of the mites in Britain are pyrethroid-tolerant to some extent.

At the beginning of the twenty-first century we still have no completely effective treatment that is safe for bees and humans alike and *Varroa* mites, along with their associated viruses, are decimating our bees. Continued treatment with chemicals to which mites can develop immunity is counter-productive, as we are simply breeding tougher mites by default.

It is interesting to note that, while agri-chemical corporations, such as Bayer, were selling pyrethroids to beekeepers as mite treatments, they were simultaneously selling them to farmers to spray onto crops. It would hardly be surprising if these low-level applications onto pollen-bearing crops turn out to have contributed to the rapid build-up of pyrethroid resistance among *Varroa* populations in our hives.

Across the Atlantic, where honeybees were unknown before settlers introduced them in the 17th century, central America and the southern states of the USA are being colonised by the so-called 'Africanized' honeybee (or AHB), also known – with some justification – as the 'killer bee', due to its unpleasant habit of mounting unprovoked mass attacks on humans and livestock, often resulting in the death of its victims from multiple stings, often many thousands at a time. The AHB is the direct result of an unfortunate experiment in cross-breeding, which escaped into the wild in Brazil. I am indebted to Marty Hardison for the following account:

> *In 1956 the geneticist Warich Estevam Kerr imported some queens from Africa. A year later his bees were mysteriously released. We will*

probably never know the actual circumstances, but Mr. Kerr was not only a scientist, he was also a highly respected human rights advocate. His criticism of the mistreatment of Brazilians limited the repressive actions of the military government.
In 1964 a smear campaign was launched against Mr. Kerr in the press. The bees he was working with were called "abelhas assassins." This label - which literally means assassin bees - was badly translated by time Magazine in their September 24th, 1965 edition as "killer bees." The title caught the fancy of the American press and Hollywood. The bees have been given a lot of hype and have caused some problems. But they don't attack without provocation: they just defend their colony aggressively. You don't want them in your yard. But they are not as fatally dangerous as bathtubs. I have worked with several colonies of the hybrid Africanized bees down in Texas. They are not as much fun to work with as our Europeans but neither are they impossible.

Interestingly, according to some authorities,[18] the AHB seem to be somewhat more tolerant of *Varroa* than our 'domesticated' varieties, possibly because it has been largely left alone by beekeepers.
In the summer of 2007, the news was full of yet another disaster to befall the honeybee: the so-called 'Colony Collapse Disorder' (CCD), which has decimated the North American beekeeping industry and seems also to be affecting Europe to some extent. Various enquiries into the cause of CCD are under way, with some beekeepers pointing the finger at the increasingly widespread use of GM crops, pesticides like Bayer's Imidacloprid (banned in some European countries but still used in Britain and the USA) and a general decline in overall bee health caused by the long-term stresses of being farmed on an inappropriately commercial scale. The latest explanation of CCD is that it is caused by a *Nosema ceranae,* previously associated only with the Asian bee, *Apis cerana.* Symptoms similar to CCD were first described as long ago as 1915,[19] when a particularly wet spring caused many losses in the USA.

Nosema apis is also associated with cool, damp conditions and stress:

Nosema appears to be highest and have the most negative impact on queens and package bees following shipment, and colonies in the spring if one or more other maladies are affecting them.[20]

Nosema apis and, perhaps, *Nosema ceranae,* may be the biggest, unreported killers of bee colonies, due to the lack of visible symptoms.

---

18 *Varroa*-tolerant Italian honey bees introduced from Brazil were not more efficient in defending themselves against the mite *Varroa destructor* than Carniolan bees in Germany. M.H. Corrêa-Marques, D. De Jong, P. Rosenkranz and L.S. Gonçalves, Departamento de Genética, Faculdade de Medicina, Universidade de São Paulo (USP), 14049-900 Ribeirão Preto, SP, Brasil

19 According to Dr. James E. Tew, State Specialist, Beekeeping, The Ohio State University

N. apis is thought to be present in 'background' quantities in virtually all hives, kept in check by the bees' immune systems until they are subjected to environmental stressors, such as being repeatedly disturbed or subjected to damp, cold conditions.

A benevolent view would be that all these unfortunate events came about as a result of the perfectly understandable but misguided desire to obtain, breed and deploy a 'better' bee - meaning, of course, better for human purposes. They were the side effects of our arrogant conviction that we can always 'improve' upon nature, and that all of nature exists only to serve our needs. We can go back to the Rev. Langstroth and find the root of the misguided attitude of pre-Darwinian beekeepers in a paragraph written six or seven years before Darwin published *On the Origin of Species* in 1859:

> *The Creator intended the bee for the comfort of man, as truly as he did the horse or the cow. In the early ages of the world, indeed until very recently, honey was almost the only natural sweet; and the promise of "a land flowing with milk and honey," had then a significance, the full force of which it is difficult for us to realize. The honey bee was, therefore, created not merely with the ability to store up its delicious nectar for its own use, but with certain properties which fitted it to be domesticated, and to labour for man, and without which, he would no more have been able to subject it to his control, than to make a useful beast of burden of a lion or a tiger.*[21]

That we have succeeded in many ways to bend nature to our will is self-evident. Take, for example, the Holstein-Friesian cow, whose milk production has been vastly and continuously increased by controlled breeding, or the truly spectacular modern racehorse, or the fast-maturing battery chicken.

Genetic engineering is the most recent manifestation of this human-centred, controlling attitude towards the rest of the nature.

> Exclusively yield-oriented cultivation ... brings the natural imbalance between plants and animals to the extreme and beyond, e.g. in cows which drag their overdeveloped milk udders over the ground with difficulty or the corn which can produce its ears only with the aid of chemical stem shortening agents. Genetic engineering takes this violation of nature a step further. By tampering with the nucleus of the cell, the plant is forced to make a fundamental change to its metabolism and creative potential, solely to serve financial interests and without any

20 James C. Bach, WSDA State Apiarist, Yakima WA, USA 1998

21 The Hive and the Honey Bee, 1853

appreciation for the essence of the plant.[22]

But surely – I hear you interject - what we have done for the cow, the horse and the dog can be applied to the honeybee, without necessarily incurring such penalties?

Indeed, by careful selection and controlled crossing we can achieve – at least temporarily - increased yields of honey. As Brother Adam demonstrated, we can likewise select for docility, disease resistance and over-wintering ability. We can, perhaps, reduce the swarming tendency, increase calmness on the combs during inspections and even – at least in theory - select for the ability to tolerate or attack mites. But if our management, medication and handling techniques continue to cause the bees undue stress and our demands on them continue to grow, they will inevitably continue to suffer, to decline in numbers and to succumb to more and more diseases and pests.

And we should always remember that, in matters of evolution, nature will select for the ability to adapt and survive, not for maximum convenience to mankind.

It is not in man's nature to be content with what he has. We see a creature that has evolved over countless millennia to thrive in a range of climates from tropical Africa to the Siberian tundra, so subtly adaptable that it can develop multiple, local ecotypes within a country as small as England, so flexible that it can live contentedly within a hollow log, a chimney or a gap in a wall and we want to impose our criteria on it: to make it behave as we desire and to produce food not only for itself but for us as well.

When beekeeping was largely the preserve of monks and peasant farmers and feral swarms were plentiful, this attitude was less prevalent and in any case, due to the limited scale of individual enterprises, did little damage. However, once mass production of hives, frames and foundation became possible, beekeeping on a commercial scale was an inevitable development. A century or so later, with the ready availability of lifting and trucking machinery, businesses comprising many thousands of hives are not uncommon and their potential profoundly to influence the health and welfare of the bee population at large is enormous.

Marty Hardison, a more experienced top bar beekeeper than myself, and a valued contributor to this book, put it this way:

> *By employing a system of migratory beekeeping, which requires transporting large quantities of hives over great distances to make optimal use of seasonal changes, we have enabled the*

---

22 From promotional material by the German company Sonett, manufacturers of environmentally benign cleaning products.

problems of isolated regions to be the problems of all. If this were not the case, American beekeepers wouldn't be dealing with parasitic mites from the Philippines, aggressive bees from Africa, or a brood disease from Europe. Neither would there be concern for the contamination of honey from the very chemicals developed to combat these problems.[23]

THE PESTICIDE THREAT

Because bees live in large colonies and gather their food from a wide area, they are particularly liable to accumulate in their hives concentrations of toxins that are unlikely to occur within the homes of any other species. This puts them in the uniquely vulnerable position of being the 'canary in the coal mine' as regards the overall toxicity of our countryside and thus our food supply.

In other words, when bees die through pesticide poisoning, it should ring alarm bells throughout the agricultural system that we ignore at our peril.

A study by Penn State researchers, presented in August 2008 at the National American Chemical Society in Philadelphia, found at least 70 different pesticides in the hives they looked at. Unsurprisingly, they found fluvalinate and coumaphos[24] in every wax sample tested, showing that beekeepers themselves are responsible for at least some of the contamination. They also detected 70 other pesticides and metabolites of those pesticides in pollen and bees: who knew there were 70 different pesticides in use? In fact, there are many more – they were looking for up to 170 different chemicals.

Every bee tested contained at least one pesticide, and pollen averaged six pesticides with as many as 31 in a sample. I hope that shocks you. But that is just the start:

> *"We are finding fungicides that function by inhibiting the steroid metabolism in the fungal disease they target, but these chemicals also affect similar enzymes in other organisms," Said James Frazier. "These fungicides, in combination with pyrethroids and/or neonictotinoids can sometimes have a synergistic effect hundreds of time more toxic than any of the pesticides individually." The EPA only looks at acute exposure to individual pesticides, but chronic exposure may cause behavioral changes that are unmonitored. Yet, a North Carolina study found that some neonicotinoids in combination with certain fungicides, synergized to increase the toxicity of the neonicotinoid to honey bees over 1,000 fold in lab studies.* [25]

---

23 Marty Hardison, Seed and Harvest, August 1992

24 Both chemicals have been widely used to kill Varroa mites.

25 As reported by Kim Flottum, editor of Bee Culture magazine, August 18 2008

So, the profit-motivated scientists at Bayer and elsewhere, who produce these poisons, which they recommend we spread around our planet, don't bother to research the possible effects of interactions between their particular brands of pesticide and all the others already out there, which may multiply their toxicity to honey bees by *one thousand* times. And the bizarrely mis-named Environmental Protection Agency (USA) appear to simply rubber-stamp the industry's own research findings.

A major incident in Germany in May 2008, which killed well over 300,000 bees (some accounts say at least 500,000), was subsequently found to have been caused by Bayer's Clothianidin, used as a seed dressing. The Penn State research tells us:

> *In an EPA fact sheet on Clothianidin, a commonly used neonicotinoid pesticide, it was disclosed that this chemical has the potential for toxic chronic exposure to honey bees through the translocation of clothianidin residues in nectar and pollen. In honey bees the affects of this toxic chronic exposure may include lethal and/or sub-lethal effects in the larvae and reproductive effects on the queen. Documented sub lethal effects of the neonicotinoids include physiological affects that impact enzyme activity leading to impairment of olfaction memory, motor activities that impact navigation and orientation and feeding behavior, and memory impairment and brain metabolism, particularly in the area of the brain that is used for making memories. When used as a seed treatment the chemical was present, by systemic uptake in corn and sunflowers in levels high enough to pose a threat to honey bees. In a 2002 survey for pesticide residues in pollen in France, Imidacloprid was the most frequently found insecticide and was found in 49% of the 81 samples taken.*

Despite this 'smoking gun', Bayer continue to deny that their neonicotinoids pose a threat to bees.

And what is worse, the current executive of the British Bee Keepers Association have somehow been convinced by these same corporations to endorse a number of their toxic products in return for payment, and to deny that pesticides play any role in the current poor state of bee health. Instead of supporting the position of our German and French colleagues, who have suffered so much at the hands of Bayer *et al*, they have taken the side of the pesticide manufacturers. That such a venerable association, constituted to protect the interests of bees and beekeepers should be so corrupted by a handful of blinkered, misguided fools is shameful.

You can decorate your hives! This beautiful ;Hardison Hive' was painted by monks of Christ in the Desert Monastery in Abiquiu, New Mexico (photo: Marty Hardison)

THE MODERN HONEYBEE

The keeping of bees is like the direction of sunbeams.
Henry David Thoreau

The honeybee has evolved by adapting itself to its environment, such that within the species *Apis mellifera* there are a dozen or so races and countless sub-species, strains and local ecotypes, each finely attuned to conditions within a particular area. In Britain this is probably less true now than it was 100 years ago, due to the massive importation in the last century and current widespread use of Italian bees (*A. ligustica*) and a few others, introduced because they appeared to possess particularly desirable characteristics.

The native black bee, perfectly adapted to our fickle British climate and able to produce some sort of a honey harvest even in a poor summer, has survived acarine and is kept by some dedicated beekeepers who claim that, in its pure form, it is both well-tempered and productive. Here is what BIBBA[26] has to say about it:

> *It is well adapted to survive in a harsh climate. It is thrifty in its use of stores; brood rearing is reduced when the nectar flow is interrupted. It forages over longer distances than the Italian bee and can make better use of meagre food resources. It will be observed foraging both earlier and later than A.m.ligustica, and will fly in dull and drizzly weather which would keep Italian bees indoors. It may also be that mating can take place at lower temperatures than in the case of the southern races. Although less prolific than Italians, the workers live longer and there is a higher ratio of foraging bees to hive bees. The wintering capabilities of the Dark bee are excellent; although colony size is at all times moderate, and the winter cluster is small, heat is conserved by the tightness of the cluster and the large bodies and long overhair of the bees. The "winter" bees of the northern race have the ability to retain faeces in the gut for long periods, due apparently to a greater production of catalase by the rectal gland in Autumn. They are thus less dependent on cleaning flights. They are also less likely to be lured out of the hive by bright winter sunshine than Italian bees.*

The Buckfast Bee, by comparison, is a docile, productive and reasonably thrifty bee[27], but has to be maintained over time by carefully controlled breeding or it will out-cross with other strains, often resulting in very nasty offspring. True Buckfast queens are now only obtainable

---

26 Bee Improvement and Bee Breeders' Association. BIBBA was founded in 1964 for the conservation, restoration, study, selection and improvement of the native and near-native honeybees of Britain and Ireland.

27 Thrifty in this sense means low consumption of winter stores.

from a few specialist European breeders. Buckfast Abbey no longer uses Bro. Adam's breeding system.

Such feral bees as still occur in Britain tend to be random crosses between imported and native varieties and as such are often unpleasantly defensive, especially when their honey is at stake. However, these feral colonies need to be watched, as those among them that thrive across several seasons may well hold the genetic key to the long-term survival of the species.

Due to imports of foreign stock over many years and with the exception of small areas where the British Black is – it is claimed - kept pure, most bees in Britain must now be considered to be mongrels and we have to deal with them as best we can, selecting those with desirable habits and qualities for breeding, inasmuch as we have control over such matters. Above all, we must aim to select for reasonable docility[28], as nothing will make us more unpopular more quickly among the general urban public than overly-defensive bees. You can be sure that, if your hives are anywhere near the site of a stinging incident, you will get the blame, even if - as is usually the case - the victim brought it on themselves by trying to swat the hapless bee.

In the USA, the picture is complicated further by the apparently unstoppable spread of the 'Africanized' bee – although winters in the northern states may keep it at bay, 'global warming' notwithstanding.

In Europe, the so-called Carniolan bee (*A. mellifera carnica*) has found much favour, especially in Slovenia and Austria and parts of Germany, where, I believe, it is the only bee allowed to be hived.

Given the relative ease and speed of travel these days and the propensity of some beekeepers to import and export – legally or otherwise – whatever strain of bee takes their fancy, the fact of the matter is that most of us will have to learn to deal with whatever mongrelized honeybees should fly in our direction: such is the challenge of 21st century beekeeping.

---

28 And I think we should allow feral bees to be as defensive as they need to be in order to maintain other genetically transmitted characteristics that may be hidden alongside the 'defensive' gene.

THE MODERN BEEHIVE

That which is not good for the bee-hive cannot be good for the bees.

Marcus Aurelius

Just about all the beehives in common use in the UK, Europe and the USA today are similar, differing only in details and dimensions. They consist of rectangular wooden boxes containing a varying number of removable wooden frames, a floor of some kind and a roof to keep the rain out. Other wooden boxes, called 'supers', with (usually) smaller frames are stacked on top to collect the honey crop. As such, they have changed barely at all from the first 'modern' hives constructed in the middle years of the nineteenth century. Arguments have raged to and fro about the ideal length, height and width, the optimum number of frames for this or that bee and innumerable other details, but we are really using pretty much the same hives that our great, great grandfathers used – at least, those among them who were persuaded away from straw skeps.

There are two possible reasons why something as functional as a beehive should remain virtually unchanged for 150 years, while all around us the engineered world has, in almost every other respect, changed utterly. Either it is perfectly suited to the job and cannot be improved upon, or its use has become so ingrained in habit and tradition that nobody has bothered to question whether or how it could be improved. In this case, I think a little of both applies: in some ways, the box-and-frame hive is reasonably well-suited to the job – at least from the beekeeper's point of view. It is a simple matter to lift individual frames out of the hive to see what the bees are doing and - if you are fit and have a strong back - it is relatively easy to remove the honey crop.

From the point of view of the bees, however, it has several important disadvantages:

- The frames are rectangular, usually wider than they are high, while bees naturally build comb in deep, catenary curves - the shape made by a chain or rope suspended by its ends.

- The use of pre-formed, worker-cell size foundation forces bees to build comb according to our requirements, not theirs[29]. They prefer to adjust the size of their worker cells according to season

---

29 A couple of years ago, I took some measurements from a conventional hive, and found that the brood foundation we currently use is about 5.65 mm, while a piece of 'free' comb (comb that had been built in a space inadvertently left by a bee-keeper) was between 4.9 and 5.1 mm across the flats of each cell. This must suggest that the bees have their own ideas about cell size.

and build drone cells according to how many males they choose to raise.

- They like to build queen cells around the edges of their comb, which is difficult if the foundation covers the full width and depth of the frame.

- They prefer to space their honey storage combs slightly wider apart than their brood frames, which is impossible if all frames are equally spaced.

A short frame placed in a British National honey super was left with a gap underneath it and the bees took advantage of the opportunity to build comb according to their own needs. Inevitably, this resulted in much drone comb being built, as their natural inclination to do so had been repressed by the use of worker-pattern foundation. This kind of a mess is never seen in natural comb.

- They prefer to live in cavities with plenty of space below their combs, while modern hives have only a small space – as little as a single bee-space - between the bottoms of the frames and the floor.

- And the very feature that make this arrangement most suitable for beekeepers – the fact that frames are movable and removable – spells disaster for bees if their caretaker chooses – as far too many do - to re-arrange their nest according to his whim, careless or ignorant of the needs of the bees.

In fact, most hives are also less than ideal for beekeepers:

- When the lid and inner cover are removed, the whole colony is exposed at once, causing a sudden temperature drop and an instant, mass protest. The beekeeper tries to silence this revolt by applying liberal doses of smoke, which, as often as not, aggravates the bees rather than subduing them, with painful and disruptive consequences.
- Frames are made to precise dimensions, which means that they must be purchased - at no small cost - from manufacturers equipped with expensive, precision machinery, and laboriously assembled with hammer and pins. They are easily damaged by rough handling and are difficult to clean thoroughly.
- Foundation wax also has to be bought in - as precision mills cost a king's ransom – and fitted carefully into the frames with more pins and fine, zig-zag wire reinforcement, close to which bees often refuse to build comb.
- The wax used for making foundation will contain a random mix of all the lipophilic chemicals that previous beekeepers have chosen to apply, as it is bought in by the millers from whoever cares to sell it to them. (When you consider that some irresponsible beekeepers routinely use organo-phosphates and antibiotics, you may not be so keen to handle it yourself, let alone give it to your bees.)
- Then, when it comes to harvest time, we have the problem of weight. A full 'super'[30] of honey can weigh between thirty and sixty pounds, depending on the type of hive and number of frames. Not surprisingly, hernias and chronic back pain are commonplace among commercial beekeepers and many people, especially women, are put off even hobby-scale beekeeping by this consideration alone.

How is it then, that after one and a half centuries of 'modern' beekeeping, we are still using equipment that has so many obvious drawbacks? Are beekeepers such remarkably conservative creatures that they cannot bear to be parted from that which they have always known? Are they so bereft of imagination that they cannot develop something more suitable for the job?

The truth is that many attempts have been made to 'improve' the design of beehives, but in almost every case they have taken one feature as given and unalterable – the holy frame - and along with it, the use of

30 The top boxes that usually contain smaller frames that the bees fill with honey.

wax foundation.

While the invention of the movable frame is commonly regarded as the greatest ever single step forward in beekeeping, it also locked into the minds of the Victorian beekeeper the notion that it was desirable – even necessary – to bend the behavior of the honeybee to the will of man; to force this wild creature to work according to the conditions they chose to impose upon it, rather than let it do things in its own particular and variable manner. This one step, I believe, sealed the fate of the bee, which has done its best ever since to adapt to this imposed regime only because we have given it no real choice. Since frames and foundation have been the unquestioned, dominant paradigm in beekeeping, perpetuated by beekeeping associations throughout the western world, the health of bees has steadily declined to the point where they are now in real trouble.

Casting aspersions on Langstroth's movable frame is bound to get me into trouble in beekeeping circles. I have no doubt that many older beekeepers who have read this far will, about at this point, give up in disgust. How dare this newcomer, with only a few years' experience, dare to question one of the most holy laws, given to us by the great Langstroth, Root and Dadant and passed down through the generations... heresy... etc...

However, I have never been one to shy away from an argument. Neither have I ever been able to do anything for very long without asking a lot of questions, so I am happy to bear whatever wrath and ridicule should come my way if this prompts a few people to take a second look at this issue and at least consider that there might be other ways to skin this particular cat.

The author with a top bar comb

BAREFOOT BEEKEEPING: A NEW APPROACH

Without husbandry, "soil science" too easily ignores the community of creatures that live in and from, that make and are made by, the soil. Similarly, "animal science" without husbandry forgets, almost as a requirement, the sympathy by which we recognize ourselves as fellow creatures of the animals.

Wendell Berry

Here we come to the principal differences between 'barefoot' beekeepers and their 'high tech' contemporaries: *the 'barefoot' or 'natural' beekeeper will aim to work with the natural impulses and habits of the bees, respecting the integrity of the brood chamber, leaving them ample honey stores over winter and generally arranging things in order to cause their bees as little stress and disturbance as possible.*

That is not to say that 'barefoot' beekeeping is an entirely 'hands off' system, or one that thrives on neglect – just the opposite, in fact. More time is spent observing the bees and some operations – at least in the horizontal TBH - may need to be performed a little more often: honey harvesting, for example, is likely to be done by taking smaller amounts over a period of weeks or months, rather than the typical all-at-once, smash-and-grab raid practised by all commercial beekeepers and most amateurs.

Barefoot beekeeping is about simplicity of equipment and of method. Whereas 'modern' (i.e. essentially 19th century) beekeeping, as practised in the early 21st century, requires machine-made wooden frames, queen excluders, manufactured wax foundation, centrifugal extractors, settling tanks, pumps and hoses and a lot of storage space in which to keep all the boxes, hives, supers, spare frames and other stuff, the barefoot beekeeper requires only one or more of the simple and versatile hives I shall describe later and a sharp knife. In place of a smoker, you can use a simple, hand-held plant spray containing water and, perhaps, a few drops of a mixture of essential oils or cider vinegar - which I will also tell you about later. And if you keep docile bees, you will use it only sparingly.[31]

Barefoot beekeepers do not aim to extract every possible drop of honey from a hive. They respect the bees' need to eat their own stores -

---

31 No, bees don't much like to be wet, but they can cope with it, as you will see if you ever find a swarm that has had to weather a shower. Neither do they like smoke - in fact it terrifies them - and this is the reason beekeepers have used it for centuries to 'pacify' them while doing stuff in the hive. What smoke really does is to trigger the bees' emergency response, which is to fill themselves with honey in preparation for a fast getaway. The problem with that, of course, is that it severely disrupts the activities of the hive for some time, while they work out that the end of their world is not quite as

especially over the winter – and regard sugar syrup as a much inferior supplement to be given only when, due to prolonged bad weather or other causes, bees are short of their own food.

I am not - let it be clear - advocating a return to the old, destructive methods of honey harvesting from skeps; nor am I suggesting that monks and farmers are the only people fit to keep bees. I do believe that a mutually beneficial and sustainable relationship with our bees must be based on a truly holistic approach: we need to learn more about how the colony works as a complete, living entity and the manifold ways in which it interacts with its environment, with us and with other living things. For too long we have been locked into an old-fashioned, reductionist approach, dealing with bees as if they were mere machines created solely for our benefit, instead of highly-evolved, wild creatures, with which we are privileged to work.

In my Introduction, I gave the three basic principles of 'barefoot beekeeping'. Now we can take a look a these principles in more detail.

1. Interference in the natural lives of the bees is kept to a minimum.

The principle of minimum interference is not a licence for neglect, but rather an encouragement to 'leave well alone'.

The British Bee Keepers' Association advocates full colony inspections every 7-14 days throughout the season, on the grounds that we should at all times know the status of our colonies in terms of diseases, parasites, queen-rightness, available space and food stores. I (and many others) say that you can gain a quite accurate picture of all of these aspects without routinely removing every comb from the hive, which creates considerable disruption and needless stress for the bees. I will go into detail on this later.

At the other extreme, advocates of the Abbé Warré hive (a vertical top bar hive; see later) leave the bees to their own devices from spring right through to the end of the season. I suspect that bee inspectors may not be entirely happy with this 'extreme' form of let-alone beekeeping, but it certainly takes our 'minimum interference' principle to its logical end, short of abandoning the concept of 'keeping bees'. Indeed, I think there is an excellent case for setting up a network 'conservation' hives, that

---

imminent as they first supposed and get busy tidying up the mess they have made of their honey stores. It also takes a while for the smoke to disperse and the hive atmosphere - heat and smell - to return to normal.

are not touched at all, and in which bees may have a chance to build up their genetic diversity once again and provide swarms for re-populating our hives.

2. Nothing is put into the hive that is known to be, or likely to be harmful either to the bees, to us or to the wider environment and nothing is taken out that the bees cannot afford to lose.

I have no scientific training beyond 'A' level, so I simply state as a matter of opinion that the way forward in the control and treatment of bee diseases and parasites lies not in the use of synthetic chemicals, but in selective breeding from naturally-surviving stock and the use of bee-friendly hives and techniques, along with natural-sized cells built by bees to their own design.

Our recent experience with *Varroa* has demonstrated that this pest can develop immunity to chemical treatments in a very short time. By using such an approach, we are simply helping along the evolution of the mite by selecting for immunity to treatments: those mites that survive our assault go on to breed with other survivors, carrying their immunity as a genetic trait. So now we have mites that are harder to destroy, Bayer will, no doubt, come up with another chemical 'cure' and the cycle will begin again. In my opinion, this is a dead end policy that bolsters the obscenely large profits of the agri-chemical industry while making the problem worse for us and the bees in the longer term.

The barefoot beekeeper has no use for synthetic chemicals, relying instead on creating the optimum conditions for the bees' health and well-being. If a treatment is necessary for the bees' survival, he looks for the natural medicine that causes least harm.

Harvesting of honey is on the basis of sharing with the bees, rather than robbing them blind. The barefoot beekeeper aims to leave the bees ample honey to get them through the winter, only feeding sugar when absolutely necessary to supplement their stores.

Considerable argument has raged to and fro about the use of sugar as a feedstuff for bees. Some say that it is just as good as honey, while others claim it to be positively detrimental. I believe that bees need honey, as that is what they make for themselves. Of course, they turn sugar syrup into something as close to honey as they can, but it seems likely that it will always be missing that tiny fraction of minerals and vitamins[32] that, on the scale of their digestive systems, make a vital

---

32 Also antibacterial and antimycotic peptides and enzymes (for example in Prof Jürgen Tautz

difference to their overall health.

On this basis, I always make sure that my bees have honey for their winter feed and if that means I have less honey on my toast, then so be it.

A good start. A well-laid-out comb from a recently-hived swarm.

3. The bees know what they are doing: our job is to listen to them and provide the optimum conditions for their well-being.

This was, according to those who knew him, one of Brother Adam's favourite sayings: *"Listen to the bees and let them guide you."*

In the literal sense, much can be learned by the attentive beekeeper by listening to the sounds the bees make and learning to differentiate their meanings. The difference in both pitch and volume between the wing notes of a bee in attacking mode and that of one who is merely curious

---

(BEEgroup, Biozentrum Uni. Würzburg) *Phänomene Honigbiene* (2007) on p176.

is one of the first and most useful distinctions. Then there is the brief 'roar' when a hive is tapped with a knuckle, which gives you information about the number of inhabitants and their general state of alertness. If the roar continues beyond a second or two and develops into an overall hubbub, this may be a clue that the colony may be queen-less.

A healthy hive of docile bees has a characteristic, contented hum, that is the most pleasant and relaxing sound the beekeeper will hear. Of course, Bro. Adam was also indicating that we should pay attention to the needs of the bees and plan our work accordingly and not by our own clocks.

I believe that there is much more to be learned by both literal and metaphorical 'listening': we just need to develop our perception by spending more time observing the bees and less getting in their way.

Skeps still in use today on the Luneberg Heath in Germany (photo: John Phipps)

A contributor to the Biobees forum extended this idea of 'listening' to bees across other senses:

> *I do that. Sometimes with the a stethoscope and sometimes just pressing my ear against the hive. I've done it since I was a kid and the bees were my dad's, so I've got a lot of bee-listening experience. I know a fellow who had a laying-worker hive in his busy, packed bee-yard and I surprised myself by picking it out from several yards away, as he was leading us over to have a peer at it. It sang a different note.*
>
> *You can definitely tell between 'we have nothing to do' and 'we are happy and making honey' and 'we are queen less or have a disease or something made us angry recently' and 'we are about to swarm.'*

With the 'scope you can locate the cluster's position in the hive in the winter without opening it, and that'll give you a guess how much of their stores they have eaten without you needing to open the hive.

Another fun sensory beekeeping experience is to try to smell the breath off your hives. Each will have its unique 'body odor' and you may also find you can tell if honey is being made, and they have much brood. You should enjoy an animal-fur smell reminiscent of a clean mink coat, plus a hint of lemon-grass (nasonov pheromone) plus a teak-hinting-vanilla woody scent (this is the smell of queen-rightness, queen mandibular pheromone and retinue pheromone are chemically similar to the benzoin perfume-fixative that this smell somewhat resembles). You are of course familiar with the floral-acidy-sweet smell of honey-production, and the resin-smells of your local propolis. The presence of a lot of brood adds a milk/ cream odor to the mix. A healthy productive hive smells wonderful. Aside from being enjoyable, smelling the stream of ventilated air from the hives is also good practice because if you take note of the normal smell you will immediately notice if there is mould, wax-moth, or foulbrood.[33]

This is an eloquent invitation to use our sensory apparatus more fully, to extend and deepen our awareness of and our relationship with our bees.

I have confined myself to sketching out these three principles, as I believe that, if they are assimilated as the foundation of our beekeeping practice, we do not need a 'Book of Rules' to answer every little question that arises. No such book could cover every eventuality and condition, but a set of sound principles can remind us where to look for the answer.

33 Contributed by Kale Kevan

TOWARDS A BETTER BEEHIVE

You never change things by fighting the existing reality. To change something, build a new model that makes the existing model obsolete.
R. Buckminster Fuller

In discussing the ideal hive in which to keep bees, it is useful to return to first principles: what kind of home do bees prefer and how do they manage it themselves, in the absence of a beekeeper?

Here is a summary of the biggest natural hive study I know of:

> *The natural honey bee nest was studied in detail to better understand the honey bee's natural living conditions. To describe the nest site we made external observations on 39 nests in hollow trees. We collected and dissected 21 of these tree nests to describe the nest architecture. No one tree genus strongly predominates among bee trees. Nest cavities are vertically elongate and approximately cylindrical. Most are 30 to 60 litres in volume and at the base of trees. Nest entrances tend to be small, 10 to 40 cm2, and at the nest bottom. Rough bark outside the entrance is often smoothed by the bees. Inside the nest, a thin layer of hardened plant resins (propolis) coats the cavity walls. Combs are fastened to the walls along their tops and sides, but bees leave small passageways along the comb edges. The basic nest organization is honey storage above, brood nest below, and pollen storage in between. Associated with this arrangement are differences in comb structure. Compared to combs used for honey storage, combs of the brood nest are generally darker and more uniform in width and in cell form. Drone comb is located on the brood nest's periphery. Comparisons among Apis nests indicate the advanced characters in Apis mellifera nests arose in response to Apis mellifera's adoption of tree cavities for nest sites.*[34]

So bees seem to prefer cylinders of 30-60 litres capacity; they are not fussy about the species of tree; they prefer low, smooth entrances; they like to coat the inside with propolis; they fix comb at the top and sides and leave passageways; they vary comb structure according to function and season.

From other sources, including simple observation, we also know that bees vary their cell sizes according to season and relative position in the nest; the main objection to the use of foundation[35], which attempts (rarely successfully) to force the bees to build their combs by our one-size-fits-all rule – the product of an inflexible, mass-production mentality.

We could say that bees prefer tall cylinders because that is how trees

---

34 Seeley & Morse, 1976

35 Thin sheets of beeswax embossed with a hexagonal pattern on which bees start building comb

grow. Yet they thrive in horizontal cylinders equally well when available, as can be seen in the suspended log hives common in the forested parts of Africa. This is just as well, as managing a vertical, tree-shaped hive sustainably (i.e. non-destructively) for honey would be very difficult indeed if one were to consider the near-impossible task of inspecting and removing an array of long, tall combs. To some extent, this has been addressed by the Abbé Warré Hive (see below).

Bearing these considerations in mind and our general knowledge of bees and beekeeping, the characteristics of an ideal hive would include key features of benefit to the bees as well as some that mainly benefit the beekeeper, while causing as little inconvenience as possible to the bees. Specifically, it would give bees the freedom to build comb according to their needs and perceptions and not force them into a pre-ordained pattern. It would offer free, untrammelled access throughout to all inhabitants. It would be provided with an accessible entrance that is big enough for free traffic movement and small enough to defend against predators. It would have provision for removing combs when required for inspection, removal to another hive, or harvesting honey. It would keep out the weather, particularly in winter, and have sufficient insulation to prevent condensation, which may chill and injure the cluster and render them susceptible to *Nosema.*

Any of our modern, movable-frame hives could be made to provide most of the above features, simply by ceasing the use of foundation and removing obstructions like queen excluders, but they would still be far from ideal, as they are not constructed to be used in this way. They are too complicated and rely on expensive, precision-made parts, which are fragile and difficult to keep clean. They are usually made of wood that is too thin to deter condensation, and they have little or no provision for extra winter insulation. They contain a multitude of awkward an inaccessible corners and joints: all but impossible to clean and ideal hiding places for wax moth larvae, small hive beetles and myriad other nuisances.

Luckily, there is a better solution: do away with frames altogether and re-think the hive shape.

THE HORIZONTAL TOP BAR HIVE

In my first year of beekeeping I did some reading on hive design and discovered that frames are not, in fact, universally used. In Africa, the top bar hive is commonly employed with great success and it also has a small but faithful following in the USA and elsewhere, among pioneering beekeepers, such as Les Crowder, Marty Hardison, Dennis Murrell,

Michael Bush and others, who also prefer a simple and low- cost approach.

The typical, horizontal top bar hive (hTBH) is a long box with either vertical or sloping sides, on which are placed simple wooden bars, 1¼” to 1 3/8” (32-35mm) wide, usually with a shallow groove cut along their lower face into which a thin strip of wax is fixed. The bees build their comb as they please - using these strips as 'starters' or guides – resulting in almost as natural a formation as would be found in a hollow tree, but with the advantage – for the beekeeper - of being individually movable.

Having discovered this simple design, I immediately set about building one and found, sure enough, that the bees built perfect, vertical comb, which I could see through the viewing window I had built into one side of the box. The one drawback of this first attempt at a TBH was that, because it was relatively narrow and had vertical sides, the bees fixed their comb not only to the top bar but also to the sides of the box, making it nearly impossible to remove individual combs without rather messy breakages.

My later designs, inspired by the hTBH boxes that I found described and illustrated on various web sites, underwent a number of evolutionary mutations until I settled on a hive that is quite easy and cheap to build and simple to use, while having certain features that make it very comfortable for bees.

A simple top bar hive with detachable legs: full instructions for building this hive are freely downloadable from the author's web site. Note the side entrance holes – a key feature of this design – that give bees direct access to the gaps between combs and prevent mice getting into the hive.

THE VERTICAL TOP BAR HIVE

Some time in the early years of the 20th century, a French Abbé by the name of Ēmile Warré, decided that the many and various beehives then in use in his native country all had one serious drawback: *they caused far too much disruption to the natural lives of the bees due to the universal use of frames and foundation.*[36]

He noted that bees prefer to build their comb in a vertical plane, in a roughly cylindrical container. Not being particularly concerned with having to lift and move boxes from time to time, he considered the bees' inclinations to be the most important factor and set about designing a vertical top bar hive (vTBH), that will appeal to many as the ultimate 'leave well alone' hive.

Abbé Warré with one of his hives

36 The outspoken Abbé died the year before I was born, so I missed the opportunity to meet someone with whom I would undoubtedly have had a lot in common.

The Warré hive (or as he himself modestly called it, 'The People's Hive') is very simple in concept and construction, being essentially a series of square boxes with an internal width and breadth of 12" (30cm) and each being 8.25" (210mm) high[37]. Each box is fitted with a set of top bars, resting in rebates cut into the top of each box, each bar being 24mm wide and 9mm thick with a 12mm gap between them, giving a centre-to-centre spacing of 36mm[38]. The top surface of the bars lies just below the top edge of the box, allowing them to be stacked with bars aligned.

Starter strips of wax provide anchor points for the bees to build their comb from each top bar down to within a 'bee space' of the one below: they build comb from underneath and so cannot avoid leaving a gap as their backs begin to touch the surface of the bar below them. This, of course, prevents a tower of such boxes from becoming a single, immovable mass.

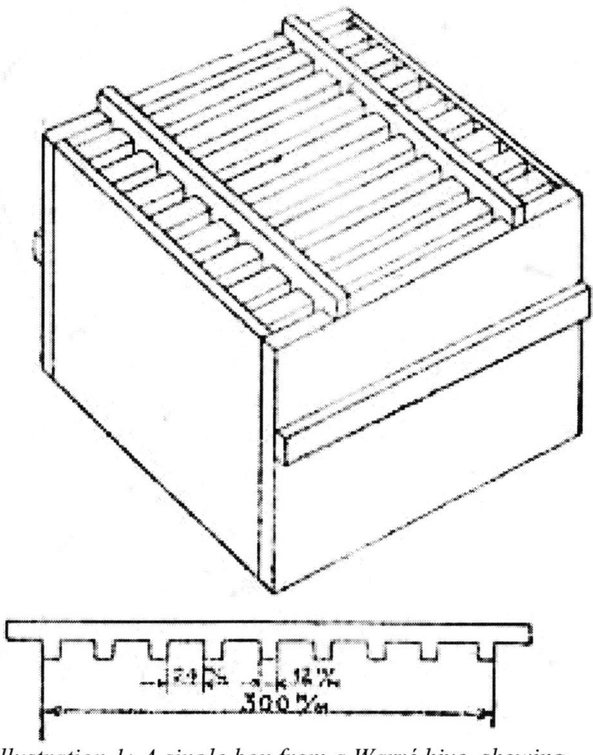

Illustration 1: A single box from a Warré hive, showing how top bars are spaced by means of a toothed spacer, which is removed once the bars have been pinned into place. In normal use, there would be three of these boxes in a stack, plus a roof and a floor.

37 This makes them considerably smaller – and therefore lighter - than any current framed hive, although the wood used is 1" (25mm) thick, which is about twice the thickness of commercial hives.

38 This is close to the 35mm spacing used by most TBH bee-keepers nowadays

The Warré hive in its original form is a 'fixed comb' hive: top bars were pinned in place and could not, therefore, easily be removed one at a time for inspection or for partial honey harvesting. This drawback – if it is to be regarded as such - could be overcome by having spacers permanently fixed in the rebates, allowing bars to be dropped into place. I suspect, however, that they would, in time, be glued into position with propolis in any case, just as the frames of a Langstroth-style hive tend to be.

The vertical TBH offers bees a reasonable approximation to a hollow tree, in which to build their comb vertically downwards, as is their instinct. They start at the very top, and having built down through one box and encountering another row of top bars, they leave a gap and simply continue downwards. Because they naturally store honey above the brood area, the top box of the stack becomes filled with honey and can be harvested all at once, while another box is placed underneath the tower so as to allow them to continue their unimpeded, downward progress. This, of course, means that boxes weighing perhaps 30lb will need to be lifted off at harvest time, while the remaining two boxes will have to be lifted to allow the additional one to be placed underneath. This is still a great deal easier than lifting a full Modified Dadant super (50 pounds or more) or a double-deep Langstroth, and could be made easier still by the use of a home-made, Roman-style, A-frame hive lifter – or an assistant.

I have built a Warré hive and populated it with a swarm, but until I have run several of them for a couple more seasons, I will refrain from casting judgement on it as a practical alternative to the horizontal top bar hive. I do think that it has considerable merit from the point of view of minimizing interference with the lives of the bees and undoubtedly is a better match for the bees' natural home – the hollow tree.

If you are now wondering whether you should construct the horizontal or the vertical form of top bar hive, I have prepared a table (see below) that may help you compare - on a purely theoretical basis - one with the other. I hope to expand on this in future editions, as I get feedback from vertical TBH users and my own experiences. I have provided no comparison of honey yields, as I know of nobody who has yet done side-by-side tests.

I am indebted to Dr David Heaf for his English translation of Abbé Warré's book, 'Beekeeping For All', which he has generously made available for free as a downloadable PDF file. Dr Heaf's material can be found at http://warre.biobees.com

	Vertical TBH[39]	*Horizontal TBH*[40]
Construction *(similar timber can be used for both types)*	Easy: minimum of 3 identical boxes, with top bars, floor, roof and stand. Rebating for top bars may require table saw or router.	Easy: one larger box, legs, top bars and roof. Only hand tools needed, although circular saw an advantage.
Portability and storage *(neither is designed for migratory beekeeping)*	Easy to move when empty; no more difficult than a framed hive when full. Some extra storage space will be required for boxes not in use.	Easy to move when empty; needs two people when full. Self-contained, so no storage problems.
Ease of use	Once set up and populated, only maintenance is removing filled upper box and replacing empty box under stack; some lifting needed, but this could be mechanically assisted.	Requires more regular attention as honey is harvested a little at a time; no heavy lifting once hive is in place.
Security	As easily removed as any stacked hive by one man and a car. Easy to hide.	Difficult to steal: would need two people and a flat bed truck or van. Harder to hide, but does not look like a regular beehive.
Location	Leave-alone style suitable for home or out apiary.	More suited to home/local use due to more frequent checks during season.
Harvesting	Usually one box at a time, by removal of top box.	Usually one or two combs at a time throughout the season.
Wintering	Usually wintered in two boxes with reduced entrance and mouse guard.	Colony wintered in same box with follower each side. Design is mouse-proof.
Inspection	Discouraged, but if top bars are not fixed, they can be removed if required.	Colony easily accessible from both ends, although disturbance should be kept to a minimum.
Feeding	Discouraged, but could be done using adapted roof and feeders designed for framed hives.	Discouraged, but when necessary can be done inside hive using simple container with float.
Swarm control	By increasing volume – adding one or more boxes below colony. Artificial swarming may be possible.	By increasing volume – adding top bars – or dividing the colony using followers. Artificial swarming easy.
Queen rearing & nucs	Not suitable, esp. if top bars are fixed: contradicts the 'leave-alone' protocol.	Swarm cells can be used: nuclei can be created within hive body using extra followers.
Brood comb renewal	Automatic: as the bees move downwards, comb is removed.	Manual: the beekeeper manages comb removal.

39 Abbé Warré hive
40 Chandler top bar hive, using side entrance and follower boards

	Vertical TBH	***Horizontal TBH***
Varroa treatment[41]	Natural cell size. Oxalic acid[42] trickling possible; powdered (icing) sugar possible from above, or below if box is lifted.	Natural cell size. Oxalic acid trickling possible by separating top bars; powdered sugar possible from below if mesh floor is used.

My first Warré hive, with A-frame stand to improve stability and raise it off the ground.

41 Our aim, as sustainable beekeepers, must be to provide the conditions in which bees can best deal with their own problems, including *Varroa*. However, benign treatments, such as powdered sugar, may be necessary in order to help the bees through the transition from frame-and-foundation hives to fully natural-cell. Once established on natural-cell comb, we aim not to interfere with their proven ability to learn how to live with a certain level of parasite infestation.

42 Oxalic acid, which occurs naturally in all plants, appears to be an effective *Varroa* treatment and the

BUILDING A HORIZONTAL TOP BAR HIVE

After much experimentation, I finally settled on a horizontal TBH design with sloping sides and side entrances as my 'standard' hive. This hive has several salient design points, perhaps the most important being that the trapezoidal shape comfortably encloses the natural shape of the comb and gives a good strength-to-volume ratio. It also virtually eliminates attachment of comb to the sides – a useful feature for the beekeeper.

A newly-built hive, showing mesh floor and two sliding follower boards. Bees will be placed in the central area, about 8 top bars added and the followers moved to fully enclose the colony. The entrance holes in this 36" hive are in the middle of the side nearest to you in the photograph. The 48" version has extra reinforcement at the end joints and two more entrance holes; one each end of the side opposite the main entrance, for use in raising nucleus colonies and artificial swarming.

This TBH has several other design features worth noting. It is simple to construct, using inexpensive – even re-cycled – materials: reclaimed, untreated pallet timber can be used, for example. All joints are glued

---

only known medication that kills *Varroa* inside brood cells, while appearing not to have adverse effects on bees if used in the recommended dilution. However, oxalic acid is extremely toxic and dangerous in undiluted form and I have read reports of the occasional queen being killed by it. For these reasons, I would rather not use it myself, but you must make up your own mind.

and screwed, which ensures strength and a long life. (If suitable machinery and skill is available, stronger, mortise or biscuit joints could be used with advantage.) When empty, the hive is light enough to be lifted quite easily by one person and it can be carried, even when occupied, by two. Because the box is bolted to the legs, it only takes a few minutes to dismantle for transport, when two or three can be carried in an ordinary car and more in an estate car or van.

This hive is at least as robust as anything currently on offer by manufacturers of conventional hives and should last many years.

The box may be made 36"-48" (900-1200mm) long (or any length you choose: I prefer longer boxes, which are heavier but give more room for bees and the beekeeper: 48" overall is now my standard) and on the outside is 18" wide by 12" deep (450 x 300mm), measured at the ends. The open bottom is covered with a plastic or coated stainless steel mesh, or may be fully closed with either a fixed or removable wooden panel.

The top bars themselves (up to 12 brood bars and as many honey bars as needed) are 17" (430mm) long, 1 1/4" (32mm) wide for the brood bars, 1 1/2" (38mm) wide for the honey bars, and about 3/4" thick. They rest on the upper edges of the sides, giving an internal width of 15" (380mm). A central groove, about 1/8" (3mm) wide and deep, is cut along the length of each bar using a circular saw and is filled with molten wax to provide a guide and anchor point for the bees to build their comb. Alternatively, a strip of thin wood can be fixed along the centre of the top bar and rubbed with wax. It is a good plan to leave about 2" (50mm) each end free of wax to discourage them from building right up to the sides.

The floor can be either solid or open mesh; the latter may be preferable in a good English summer and warmer climates in general, as it allows for plenty of ventilation and prevents the build-up of debris, while some method of closing it off may be a good plan in harsher winters if your bees are likely to exposed to strong winds. A satisfying number of *Varroa* will also drop through a mesh floor and cease to bother the bees. If you have daily access to your hives and are inclined towards statistical analysis, provision can be made for catchment trays for the purpose of counting the mite drop rate, from which deductions can be made about the level of infestation.

The outer surfaces of the hive can be coated with a 1:20 mixture of beeswax and linseed oil, melted together, well stirred in a double boiler and applied while still hot. This coating is harmless to bees and will keep the weather out most effectively without recourse to potentially toxic chemicals. If it contains some propolis as well, then so much the better - after all, Stradivarius used it to varnish his violins.

In short, this long hTBH is strong, self-contained, versatile and easy to build, even by someone with only basic woodworking skills. It is also easy to manage, as I will explain. Once the hive is in place, the heaviest lifting you will need to do is to remove the roof.

There are many other TBH designs around (see my website for links) and they all have their pros and cons, their fans and detractors. They all work – thanks to the versatility and adaptability of the honeybee.

A 'Hardison' top bar hive' in Africa (photo: Marty Hardison)

A NOTE ABOUT ENTRANCES

Reasoning from the aforementioned study, bees make their entrances to their 'wild' homes according to the formula:

$$H = V^2/90$$

where H is the area of the entrance hole in square centimetres and V is the volume of the cavity in litres.[43] So a 30 litre hive would need a $10cm^2$ entrance, a 45 litre hive would need $20cm^2$ and a 60 litre hive, $40cm^2$.

---

43 This formula was devised by my daughter, Fahran. I don't imagine that the bees actually use this formula themselves.

Natural entrances are more likely to approximate to circles than rectangles, so I have chosen to use 1" (approx. 25mm) holes rather than rectangular slots. Given that the area of a 25mm hole is very close to $5cm^2$, we can say that two 25mm holes are adequate for a small colony in a 10 litre cavity (say, a nucleus box) and that a really big hive of 60 litres capacity might have eight such holes, although in practice, three holes seems to be plenty and facilitates defence of the hive, while a nucleus colony is happy with a single entrance hole. One reason for choosing holes of this size is that they can easily be closed using champagne corks, of which there is a plentiful, free supply from restaurants and hotels, if not from the beekeeper's own cellar.

The placement of entrance holes is worthy of a little discussion. Wild/feral colonies appear to prefer their entrance to be sited at the bottom of their cavity, with clear space between it and their comb. This would make it easy for them to remove debris and suggests that top ventilation is not required when they are housed in a cavity – such as a hollow tree - that has no condensation problems due to its excellent insulation properties[44]. It also suggests that they prefer some space below their comb, which is not provided in modern hives, unless 'slatted racks' are fitted.[45]

My first TBHs had entrance holes at one end. I drilled a single hole about 2/3 of the way up the box, which seemed to work fine until the bees found their own way out through a gap between the mesh floor and the side of the hive, which I had carelessly failed fully to close. They expressed their preference for a low entrance by closing up the hole I had made almost completely with a curtain of propolis. Since then, I have used entrance holes drilled low down in the sides, which allow the incoming bees free access to several gaps between combs without having to manoeuvre around a comb face. This arrangement also makes the hive totally mouse-proof, as there is no access to the entrance other than by climbing a slippery overhang, guarded by bees.

There are several other important advantages to this arrangement, which will become apparent later, when we discuss hive management.

---

44 Commercial, wooden hives constructed from the typical thin timber used today do pose a considerable condensation problem in the damp, British climate. When I worked at a commercial bee-keeping operation, we opened many hives one spring that were black with mould on the outside and housed a dead colony. The cause of these deaths was never established, but it seems likely that condensation played a part, along with *Varroa* mites, associated viruses and perhaps, the synthetic miticides that had been used the previous season.

TOP BAR HIVE PLAN

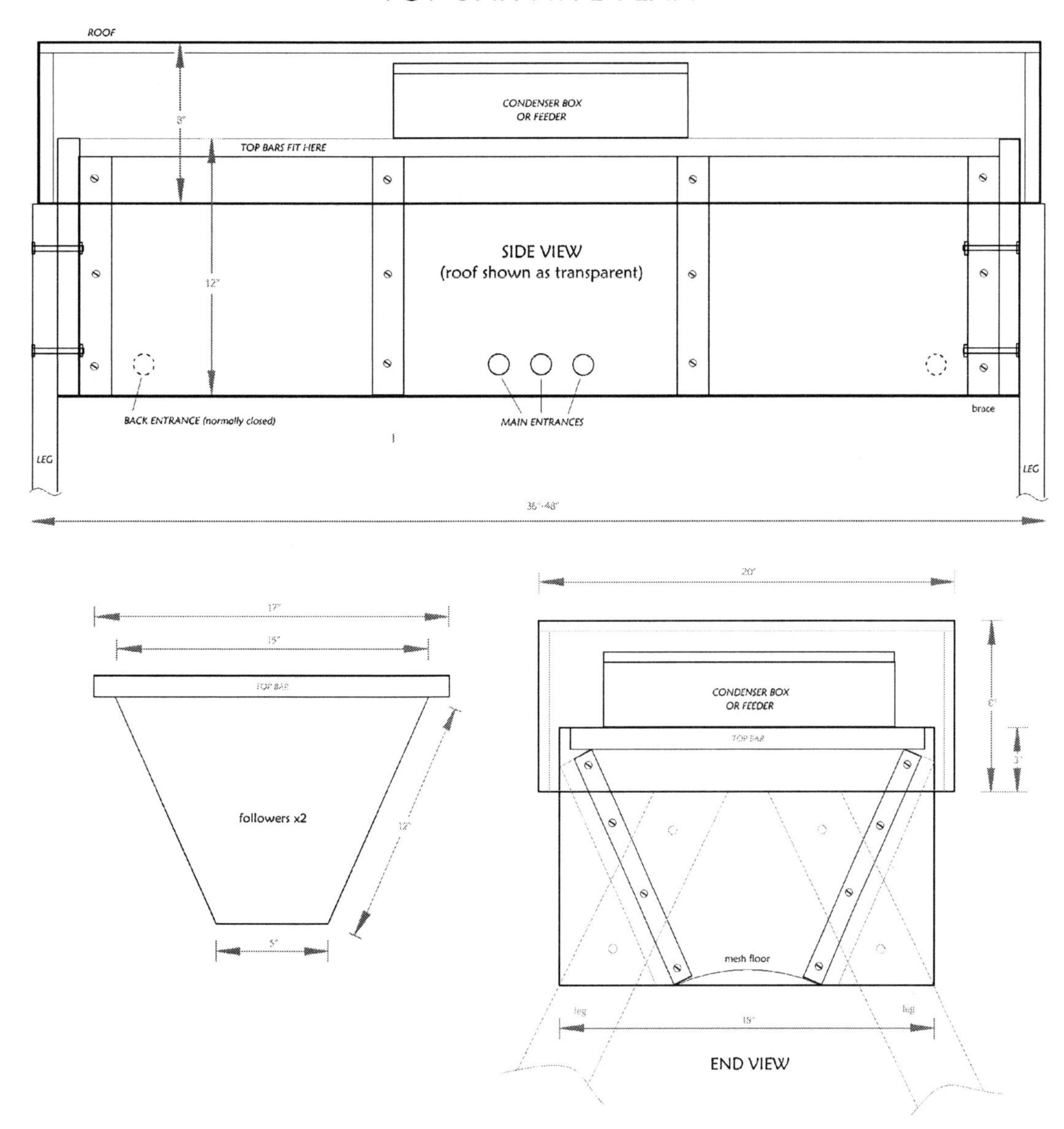

NOTES

1. The hive shown here has a simple box roof that fits over the ends and rests on the tops of the legs. If you build your hive without legs, you may want to add a 'handle' each end for the roof to rest on. If you do use legs, cut them to a length that places the top bars at a comfortable working height and fix with coach bolts for strength and to facilitate removal for transport.

2. The gap between the ends of the top bars and the inside of the roof is shown as about an inch – it does not need to be this wide, but should not be less than one quarter of an inch to avoid accidental movement of bars when removing and replacing the roof. An extra strip of wood may be added inside the roof

45 An extra box fitted below the hive, having horizontal slats fitted to correspond with the spacing of frames.

sides, to form a seal against the top bars that may deter curious wasps.

3. The 'condenser' is a simple box with a mesh floor and a board roof, made from untreated wood, containing sawdust or raw wool. It is identical to those made for the Warré hive and performs the same function: absorbing excess moisture from the air at the top of the hive, while providing effective insulation that enables the bees to maintain their desired temperature, regardless of the weather.

4. The mesh floor is shown as curved – if you use a soft, plastic mesh, nailed to the lower edge of the side panels, it will form such a curved profile and when the follower boards are fitted, it will form a bee-tight seal. If you use a metal mesh, or some other arrangement for the floor, you will need to bear in mind the profile of the followers.

5. If you make the followers first, they can be used as formers for the shape of the hive. The illustrated, step-by-step construction plans available from the biobees web site show this in detail.

6. If a top feeder is to be used, a gap must be provided for access by the bees. This can be done by means of gapped bars, or a notch may be cut in the top bar attached to one of the follower boards, which can be plugged or covered when the feeder is not in use.

7. If gapped bars are used, some form of cover cloth will need to be laid over the bars to prevent bees having access to the mesh floor of the condenser. Traditionally, this is made from a heavy cotton or hemp cloth treated with a flour and water paste to impregnate it with starch.

8. I have found that braces fixed with wood screws as shown will greatly reduce warping in the side panels.

9. Some people have fitted a hinged or removable floorboard underneath the mesh floor in order to retain as much heat as possible, especially in winter. I am coming round to the idea that a floorboard is a worthwhile addition, so long as there is enough ventilation to prevent the accumulation of moisture.

10. The dimensions shown are for guidance only and you can vary them to suit available timber and your own ideas, but I would recommend that you stay close to these measurements for your first hive unless local experience indicates otherwise.

Guards at the entrance. Landing boards are quite unnecessary: bees are just as happy either way up.

SETTING UP A TOP BAR HIVE

Like the bee, we should make our industry our amusement.
Goldsmith, Oliver

If you are setting up a horizontal TBH for the first time, this is my recommended procedure.[46]

The hive body, having been bolted to its legs (or stood on blocks, or suspended from ropes, or some other arrangement of your own choosing), should be sited on firm ground in an open, sunny spot with some shelter from the wind and away from public walkways. In Britain, some shade is acceptable, but cool, dark or damp sites should be avoided. In hotter climates, shade from the midday sun would be a good idea to prevent overheating.

If you use my TBH pattern, the bees will be offered the central section of the hive, enclosed by sliding follower boards. This has several advantages compared with trapping them against one end:

- Better winter insulation – there is an enclosed, empty space either side of the winter cluster, which can, if felt necessary, be filled with insulating material. More than one colony can be over-wintered in a large TBH, keeping each other warm.
- More control over honey extraction – honeycomb may be taken from either end.
- Easier inspections – bees can be observed working at both sides (as well as the observation window, if you incorporate one).
- Flexible artificial swarming and divisions (see later).

Everything you need for managing the hive can be kept in the hive itself. If you have several hives, you will probably want to carry your hive tool (or knife) and smoker (or spray bottle) around with you and you will probably be wearing your veil.

The purpose of the followers in my design is mainly to isolate the colony from the ends of the hive, in order to maximize winter insulation and summer cooling, as appropriate. Sloped sides also play a part in this. Side entrances also enable the beekeeper to perform splits and artificial swarming in the same box, at the same level - avoiding much of the lifting associated with framed hives.

Although I describe the entrance as central, that is true only in relation to the hive itself and not necessarily to the colony, which is usually

46 I will not cover setting up the vertical TBH here, as it is dealt with thoroughly in Abbé Warré's book. See warre.biobees.com for David Heaf's excellent translation.

offset to one side of the entrance in order to leave room at one end for the management operations mentioned. This means that the brood will mostly be found near to the entrance, while the honey will be stored towards the end of the hive. The exact arrangement of followers is at the discretion of the beekeeper, who will, of course, take direction from the bees.

The side entrance holes also, I believe, improve the bees' capacity to ventilate their hive, due to the gaps between several combs being accessible to both bees and air. Pollen- and nectar-laden bees can also gain easy access for the same reason.

A single comb lifted from a top bar hive. With care, comb can be inspected as easily as in a framed hive.

I have no doubt that end-entrance TBHs work as well, and my innovations are an attempt to improve on them and make the TBH more versatile and suitable for a wider range of climatic conditions. Only time will tell if I am on the right path.

The fact that, once the hive is in place, there is no heavy lifting to be done at all[47] and everything happens on one level, makes this style of beekeeping particularly suitable for people with mobility problems and those who do not relish shifting boxes weighing up to 50lbs at a time. Wheelchair users can arrange for these hives to be fixed at a convenient height so they can be managed from a sitting position. People who are well below or above average height can likewise arrange matters to suit themselves. The bees will not mind what height they are off the ground: they are comfortable 15' up a tree.

One of the many advantages of this horizontal hive on legs is that it is naturally mouse-proof, being well away from ground level and having an entrance that is inaccessible even to the most agile rodent. Woodpeckers, which have been known to chisel their way through standard hives, will find the extra thickness of timber a deterrent and they will find clinging to an overhang coated with slippery wax and linseed oil quite a challenge.

If there are skunks or bears in your area, you may require additional security measures. This is not yet a problem in Britain.

Perfect, new honeycomb built in the typical heart shape on a top bar in one of the author's hives

---

47 With the possible exception of a 180 degree rotation – see 'swarm control'

MANAGING A HORIZONTAL TOP BAR HIVE

Managing a TBH is easier in many ways than managing a framed hive. It is not, however, a licence to ignore the bees completely: there are certain things that must be done at certain times.

As most people start beekeeping with a swarm or a nucleus colony in the spring or early summer, let's take that event as our starting point.

A swarm settling into a temporary flowerpot hive. Note the leaves, which are still attached to the twig that was cut to remove them from a bush. The bees on the edge of the pot are fanning a scent from their Nasonov glands, which attracts other bees to re-join the colony.

Catching a swarm is an exciting introduction to beekeeping and I recommend you have a go at it at the first opportunity. Even though a large swarm can look quite menacing to the novice, bees are usually very docile and easy to handle at this stage of their development.

INTRODUCING A SWARM

If you catch a swarm, you can introduce them to your TBH quite easily.

(1) Place two follower boards about eight bars apart (or a couple less or more, depending on the size of the swarm), either side of the three central entrance holes, all of which should be closed with corks or wooden bungs.
(2) Gently shake the bees into the space between the follower boards and sprinkle or sieve a handful of icing sugar over them to give them something to take their mind off absconding, and to remove most of the phoretic mites that may have hitched a ride.
(3) Place top bars across the hive over the bees and adjust the follower boards to completely enclose them.
(4) Replace the roof and open one of the central entrances to let the bees fly. (If you are nervous about this, you can tie a long string to the cork and pull it out from a distance.)

Job done.

If you used a flowerpot hive (see illustration) to catch a swarm, they can be easily transferred to the TBH after a few days, simply by lifting them out, one bar at a time and placing the combs centrally in the same order. I would recommend this method of hiving a swarm, as it gives the bees time to establish themselves as a colony and they are less likely to abscond: the flowerpot is simply a small top bar hive.

A typical set-up for accepting a new swarm, which would be housed in the central area between two followers (marked with black dots). One entrance hole is blocked with a cork. The top bars between the followers would be removed to introduce the swarm and immediately replaced. The top bars shown on either side are spares, ready to be inserted when needed..

Bees in a new hive will go about their business with great enthusiasm, building beautiful, natural comb from the top bars, using the wax-filled grooves as guides. This will happen most times, but there is an old saying that 'bees do nothing invariably'.

If there is plenty of nectar around, after a few days the bees will be wanting more space, so choose a suitable time (early afternoon is a good time in sunny weather) and gently move the follower board furthest away from the entrance and take a look. If they are building comb within two bars of the end, add another empty bar and close it up again.

If there is not much nectar to be found when you first hive them, the bees will appreciate being fed (see below), as they need plenty of fuel for wax production.

If the flow is strong, check again in another 2-3 days and expand the nest as they demand by their building. *Listen to the bees and they will tell you what they need.* Err on the side of giving them more room and you will be amazed at how fast a strong colony can build comb.

As the comb-building worker bees do their job, the queen is busy laying and the flying bees are working hard to keep up with the demand for nectar and pollen. Another group of bees will be lining their new home with propolis, to keep disease at bay.

This period of rapid expansion will continue unabated (you can listen to them working at night if you put your ear - or a stethoscope - against the side of the hive) during spring and summer while there is plentiful food to be found.

Bees building comb: notice how they form chains, as if to measure the space.

INTRODUCING BEES FROM A PACKAGE OR NUCLEUS

Because all the bees you are likely to see offered for sale will be in the form of nucleus colonies (on frames) or what are (mostly in the USA) called 'package' bees (bees in a box with a queen held captive in a separate cage), you will have to 'convert' them to top bars in order to keep them in your new hive. Unfortunately, there is no quick and easy way I know to do this, other than a simple 'shook swarm', but that means losing all their brood and is a major upset to the bees. If conditions are good, however, they recover quickly and will make up for lost time and, in any case, losing brood will also ensure a reduced population of *Varroa* mites.

A less disruptive method for conversion from a framed nucleus colony is to build a 'conversion hive' with straight sides, designed to accommodate a standard frame. In Britain, this would require an internal width of 14 1/2" to allow 1/4" bee space[48] either side of the frames. A five-frame nucleus colony would be placed into the conversion hive, with a top bar inserted either side of the three central frames. Once these have been built out and filled with brood, they would be moved one step in towards the centre and new empty bars placed on the outer sides of them. Then the central comb would be moved out and so on until the frames would be worked towards the periphery and finally removed. The built-out top bars could then be transferred to our sloped-sided TBH, with some trimming probably being required.

However, this process is time-consuming and means disturbing and re-arranging the brood nest several times over a few weeks: far from ideal, particularly as frames have gaps either side of their top bars, which means making a special cover strip for them – another complication.

Here is another plan that is quicker to execute and requires no special hive. First, establish your nucleus colony for a week or two on a site where you want your top bar hive to be – with its entrance at a similar height from the ground and facing in the same direction as your TBH entrance will be.

Once the bees are happily flying to and from this spot, move the nucleus box several yards away and replace it with a prepared, empty TBH.

Have ready a big sewing needle – a 3" sacking needle is ideal – and some hemp or jute string threaded onto it. Place one of the follower boards flat on top of a few top bars, arranged towards one end of the TBH, or on any other handy flat surface close by.

1. Before you start this procedure, look for the queen, catch her in a tube trap[49] or gently coax her into the new hive; avoid shaking or dropping her onto a hard surface and do not touch her with your fingers, if possible.

---

48 A bee space is generally reckoned to be about 1/4"- 5/16" or 6-7mm: it is the space that bees will leave free from comb or propolis.

49 A tube trap is easy to make from a short piece of plastic tubing – see illustration.

2. Remove one of the outer frames from the nuc box - which will most likely be full of honey and pollen - and brush any bees off into the central part of the TBH using a goose feather or soft bee brush.

3. Using a sharp, serrated knife, cut through the comb (including wires, if any) around the inside the frame, removing it in one piece and placing it carefully onto the previously arranged follower board, centrally and hard up against the top bar.

4. Take your knife and trim the comb so that it matches the shape of the follower board and about 1/2" smaller on the two sloping sides and bottom edge.

5. Take your sewing needle and thread and - you may need help with this - holding the comb centrally against the underside of an empty top bar, sew it with big stitches onto the bar, without pulling too much on the string. Tie the ends off in any neat way that pleases you, so long as there are no knots on the sides of the bar.

6. Place the top bar with sewn-on comb in the TBH and repeat with the other outer comb. I suggest doing these first as they are unlikely to contain brood, giving you a couple of chances to practice your technique before handling the inner combs that will contain larvae and pupae and which need to be sewn fairly quickly to avoid chilling the delicate brood. I'm afraid you will most likely have to sew through some brood cells: avoid doing so if you can.

7. Finally, tip, tap and brush all the remaining bees out of the nuc box into the TBH, close everything up and leave the bees alone for three days. This is one time when some feeding is likely to be necessary, unless there is a nectar flow on.

The above method of conversion is not exactly easy, but it does seem to work well, as bees are reluctant to abandon brood. Flying bees take care of themselves, as they will fly to the new entrance automatically. So long as you have relatively docile bees and take care to ensure the queen is not harmed, it should all go smoothly.

N.B. It is best always to avoid touching the queen, as human smells can cause her workers to reject or even kill her: I have seen this happen even with experienced beekeepers. Use a tube trap *(see illustration)* to keep her safe whenever you have to move combs around.

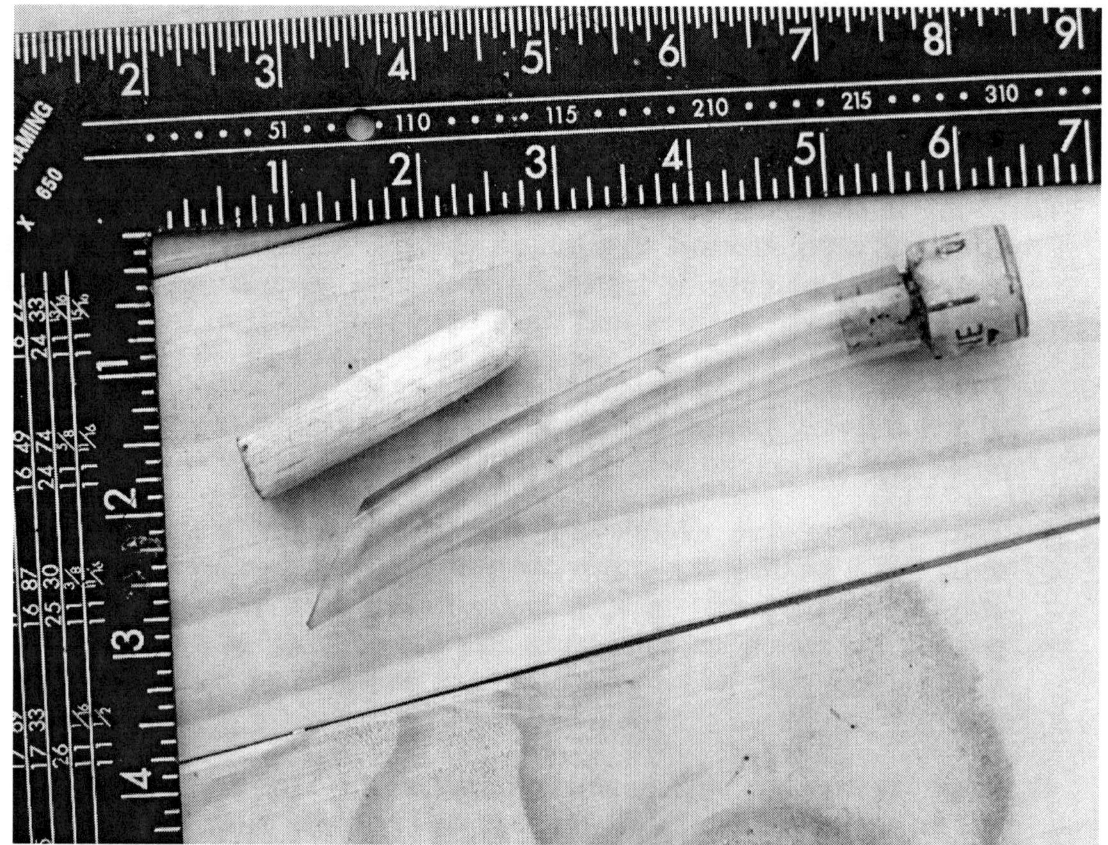

A simple queen-catcher or tube trap. The end with a sloping face is placed over the queen so that she will walk up into the tube. The wooden plug stops her backing out, while the cork bung can be removed to return her to her colony, untouched by human hand. Bees are said to have a sense of smell at least 40 times more sensitive than ours.

ROUTINE INSPECTIONS

From the time when brood rearing begins in earnest in the spring to the close of the season, occasional checks should be made to assess the condition of your colonies and to check for signs of disease and infestation.

Such checks are at least as easy to perform in horizontal top bar hives as in framed hives, so long as care is taken to avoid breaking combs at their point of attachment to the top bar. This is simply a matter of technique and can be mastered with practice. However, there is no need to dismantle the colony every week, every other week or even every month, so long as external signs of colony health are positive. *Removing bars from the brood nest is extremely disruptive and should only be done when really necessary*, such as when queenlessness or disease is suspected. The two-follower-board system allows easy access to both ends of the colony and facilitates quick inspections that the bees hardly notice and does not disturb the brood nest. You could, of course, adapt my design to include a window in the side away from the entrance, from which you could gather much information without even moving a follower board. My very first TBH incorporated such a window and I think every bee-keeper should have at least one such 'observation hive': apart from their usefulness in monitoring bee behaviour, they are a great way to get children interested in bees.

Choose a warm, sunny day in the early afternoon, when the flying bees will be out foraging and the other bees busy with their various house duties. This is the best time for a routine inspection, unless there is a thunderstorm approaching, which can make bees rather irritable. Bees will always let you know what mood they are in.

Approach your bees calmly and always work gently and slowly. If you knock and bang woodwork around as I have seen some beekeepers do, you have only yourself to blame if the bees respond in kind.

Lifting bars from the hive is done from the outside working in. Take the roof off and begin work from the end furthest from the entrance, away from the brood nest. Gently prise the follower board away from the first top bar, using a hive tool or stout knife. Move it back towards the end of the hive, to create working space. Few bees will be disturbed, as you are at this point well away from the brood nest. Beekeepers who are used to framed hives will notice that, *when working a TBH, relatively few bees are disturbed or exposed at any one time*, which means less defensive behaviour, less stress for the bees and fewer stings for the beekeeper.

If the bees seem inclined to be defensive, a couple of squirts from your water spray will calm them down. Smoke is rarely needed and if it is, you have probably chosen the wrong time to open your hive, or you have genetically bad-mannered bees, in which case re-queening may be the best option.

Combs not required to be lifted for inspection (those that are empty or contain no brood) can be pulled towards the ends of the hive, while those that need to be examined can gently be prised up, pulled away slightly and lifted vertically from the hive. If there is any sign of attachments to the side, cut them first with a sharp knife, keeping the blade against the hive wall and cutting from the bottom upwards. To avoid breakages in fresh comb, keep it vertical at all times and do not be tempted to turn it on its side, even by a few degrees.

It is important to move a top bar away from its neighbour before lifting it, so as to avoid dragging adhering bees over those on the next comb. This is even more important with a TBH than with a framed hive, as the comb is likely to be somewhat irregular in shape.

The recommended method of inspection with framed hives is systematically to remove each frame and examine both sides for signs of disease and mite infestation, for recently-laid eggs (proving the queen to have been present within the last couple of days), for multiple queen cells (in spring to mid summer – a sure sign that the colony is about to swarm) and for adequate space in the hive.

While all of these factors are important and should not be neglected, let's look at how they can all be addressed in a horizontal top bar hive while causing minimum disturbance to the bees:

1. **Checking for adequate space.** If bees are overcrowded, they are more likely to swarm. In temperate zones, this is only an issue during the build-up period in spring to early summer, so at this time you should ensure that the follower boards are moved and new bars inserted so as to stay at least one bar ahead of the expanding colony. At the busiest times, you may need to check every 2-3 days; this only takes a minute or two and need not disturb the bees.

   It is not unknown for an overcrowded colony to spill out of the entrance and for bees to be found hanging in a cluster outside the hive in hot weather. You can take it that this is a sign of their needing more space and that they are considering swarming.

   The colony will contract towards winter, or in hotter, drier regions, during the summer dearth.

2. **Checking for adequate food stores.** Bees should never be short of food, or they will suffer. So may the beekeeper: "a hungry bee is an angry bee", as Bob Marley would probably have said, had he been a beekeeper.

   The most dangerous time for bees in the temperate zone is late winter to early spring, when they may have consumed most of their winter stores and may yet be confined by bad weather. The queen is laying and there are more mouths to feed, but, if the weather is wet or cold, nectar fails to appear in quantity. This is the time that bees may need to be rescued by emergency feeding: and this situation may also arise during any extended period of drought or wet weather in summer. *You should be especially vigilant in checking for adequate stores of honey and pollen when the weather is very hot and dry, or cold or wet* for extended periods in spring and summer. We will deal with winter stores later.

3. **Checking for queen-rightness.** If, for any reason, a hive becomes queen-less, it will descend into a state of anarchy and chaos, due to absence of queen pheromones, which have a cohesive effect on the colony.

   You will, with practice, be able to spot this condition before you even open the hive: the bees will be milling about aimlessly, making a louder noise than usual; a kind of roar or loud hum that is unique to the state of queenlessness. When you open the hive, they will appear animated, though usually not over-defensive and the noise will increase and persist. You will find no brood and most likely little or no food, as they have given up foraging. If they have become queen-less within the last few days (as demonstrated by the presence of freshly-sealed brood), you can probably rescue the situation by giving them a queen from a nucleus hive, if you have one, or a swarm cell from another hive, if available. If you have no spare queens, giving them a frame of freshly laid eggs and another of honey and pollen will often work – they will usually raise a new queen and all will return to normal. If they have lost their queen more than a week ago – and especially if you find cells with multiple eggs in them (a sign of laying workers) the situation may be hopeless and all you can do is combine them with another colony – checking first for any sign of disease.

   Only experience will teach you to differentiate between a normal colony and a queen-less one and you should seek out opportunities to gain such experience by visiting other beekeepers and attending your local association meetings.

If you are new to beekeeping, I strongly suggest that you take with you an experienced beekeeper, especially the first couple of times that you carry out inspections. Take the time to observe the bees flying in and out of the entrance and discuss your observations with your colleague. Consider carefully before you open a hive, and have a clear idea why you are doing so. *The time to make contingency plans, should something unexpected be found, is before you lift the lid.*

4. **Checking for swarm cells.** During the late spring and early summer, attempted swarming is likely in some hives even in the most efficiently managed apiaries. The swarming instinct is one of the most basic functions in honeybees and is the mechanism by which they became so successful as a species. Many breeders, including Brother Adam, have attempted to breed this tendency out of them, but none, I think, have been more than partially and temporarily successful. I believe this to be a good thing: if the

bees lose this instinct completely, they will certainly perish, as it ensures their wide distribution and a chance of escaping – at least for a season or two - the consequences of modern beekeeping methods. It may also be their way of breaking free from a diseased or infested hive.

When you find swarm cells, be happy! You have an opportunity to increase your stock of bees at no cost (see below for details). On no account simply remove the cells and imagine you have solved the problem: at best, you have delayed the inevitable: at worst, you have destroyed their chosen queen.

Unwarranted and ignorant suppression of instinctive behaviour in any species can only lead to bad consequences.

5. **Checking for disease and mites.** A detailed account of the legion diseases and pests that bees are heir to is beyond the scope of this work and is well covered elsewhere. I recommend that you obtain the latest booklets (free in the UK) from your local bee inspector and study the subject in some depth – at least enough to recognize the more important diseases. Hopefully, you will never find them in your hives. Mites – the dreaded *Varroa destructor* – are now pretty much everywhere and you should certainly be able to recognize them and symptoms of their presence. Again, your local bee inspector will have a supply of the latest official literature on the subject.

If you focus your efforts on creating a healthy, supportive environment for your bees, allowing them to fulfil their natural inclinations insofar as they do not utterly negate your desire to gain some kind of a honey crop, none of the dread diseases and few of the less serious ones should bother your bees.

HANDLING COMB

Comb handling in a TBH - especially fresh comb, full of honey - is undoubtedly tricky. Here are some tips that may help:

- Before moving ANY comb, make absolutely sure that it is not attached to the sides. If in doubt, run a sharp, flexible, knife blade. held flat against the side of the hive, from bottom to top on each side. Work slowly so as to avoid damaging any bees.

- Always hold the top bar level in both planes. In other words, DO NOT TWIST IT from side to side or end to end - not even a little bit - or the comb WILL break.

- When harvesting, have a container handy in which to place cut honeycomb. A wide Tupperware or similar box with a lid works well. Ideally, have someone to help you - one holds the top bar, while the other steadies the comb from underneath and slices it off an inch or less short of the bar.

- Work slowly from one end, moving only one bar at a time. If you do have an accident, that will limit potential damage.

- Bees are better at tidying up than you are. If you make a mess and things start to go wrong, DON'T PANIC - just pick up what you can, close the hive and leave it alone for a day. Bees can shift a lot of honey very quickly if they need to.

- NEVER leave a hive with honey exposed to the open air for more than a few minutes, as the smell of honey will attract bees from other colonies and may trigger robbing. If a lot of honey is exposed in a hive, close it and reduce the entrance - especially if there are other colonies nearby.

Making a mess in a hive can certainly be disheartening, but bees cope well with disaster if left alone to do so. Put your first dropped comb down to experience and take it more slowly next time.

In temperate zones, when you open a hive - even in high summer - *the outside air is always cooler than the brood nest* (approx. 95 F), whereas in hotter places, the opposite is the case and meltdown is an ever-present danger.

Either way, opening a hive causes the bees extra sir-conditioning work, so it is always a good idea to do inspections or harvesting when the inside and outside temperature is as balanced as possible. So in the UK, the middle of a 'hot' summer day is fine for a check-up, whereas in Texas, that would be potentially disastrous.

Local conditions are a major factor in beekeeping. Adjust everything you read here, and in other books, to suit your climate.

This queen has been marked to make her easy to find. I regard this procedure as invasive and potentially dangerous to the queen, but some beekeepers find it useful.

(photo: John Phipps)

THE SIGNIFICANCE OF CELL SIZE

One of the many good reasons for using top bar hives is that bees are free to build comb to their own design, i.e. with cell sizes that vary according to season, climate, the need for drones and any other factors that the bees instinctively consider important and we know very little about. One effect of this is that – depending on the strain of bee employed - they may tend to build slightly smaller worker cells than those dictated by the pre-formed sheets of foundation used in framed hives. According to the observations of several long-time top bar beekeepers, these smaller cells appear to be less attractive to *Varroa* mites. If this turns out to be the case wherever bees are encouraged to build natural comb, then we shall have a huge clue, both as to how the mites have become ubiquitous so quickly and how best to create conditions in which the bees can best deal with them.

It seems clear to me that bees have a powerful need to build comb. It is a part of their natural life-cycle and a part of their biochemical make-up to extrude wax and to work it, and I strongly believe that they need the freedom to build it their way. If that means they build a whole range of cell sizes and raise 15% of their colony as drones, then so be it: that is what they need to do and we may never know the reason why, nor do we need to. The current pre-occupation with drone culling as part of so-called 'integrated pest management' cannot but affect the quality of queens, as many of the most important traits – including temperament - are passed down the drone line, according to the late Brother Adam and others. It would not surprise me if the many stories of poor quality queens I have heard and read about recently were caused by a local shortage of good drones.

Long-term users of natural comb hives, both top bar and framed, consistently report greatly reduced mite levels; below the danger threshold, certainly. We should consider the use of foundation-less beekeeping to at least have the potential to massively reduce the threat from this pest and, by implication, from other pests and diseases that have become dangerous through the abuse of the honeybee over the last 150 years. Further research is needed to discover the mechanism by which small/natural cells discourage mites and TBH keepers have an opportunity to be at the forefront of this research, however informally they carry it out.

If your bees originated as a swarm from another beekeeper's hive, they will in all probability have been raised on artificial foundation and the small-cell proponents claim that they need to go through one generation of free comb-building in order to return to their true size. If they are right, this means that the swarm will initially build comb, perhaps a shade smaller than that on which they were raised, while the next generation – those born from this first comb – should build truly 'natural' cells.

There is some contention about this cell-size business: some believe that it is necessary to go through several stages of reduction before bees return to their natural state; I have to say that this is not my experience: I have found bees that were born on standard foundation building free comb with an average 4.9mm cell size. Neither do I subscribe to the notion that 'small-cell' foundation is a significant improvement on 'large-cell', except, perhaps, as a temporary expedient for helping the bees deal with a *Varroa* infestation. I think this is a case where we should let the bees decide: natural cell size only, whatever dimensions they turn out.

TREATING A COLONY FOR VARROA MITES

In the transition period from foundation-raised bees to natural-cell bees, you will, in all probability, lose some colonies to *Varroa destructor* unless you use some form of treatment. As we are here concerned with keeping synthetic chemicals out of the hive, the possibilities for treatment are reduced to the use of one of the naturally-occurring miticidal acids – such as formic, lactic or oxalic – or powdered sugar. Thymol-based treatments may also be considered if handling toxic acids does not appeal, on the grounds that thymol occurs naturally in plants of the genus *Thymus*, such as common thyme. Thymol is most often applied in the form of an inert powder that has been impregnated with the active ingredient, which is placed inside the hive on something like a shallow tray. Bees absolutely detest it – as can be observed if you accidentally spill some on them – and evict it from the hive in short order, thus spreading its vapour throughout the colony. It has been shown to be quite effective when properly used.

The big problem with all of these potentially toxic treatments is getting the dose and the application right. Get either wrong and you risk killing or injuring bees (or yourself) on the one hand or failing adequately to treat on the other. If you use any of these substances, be sure to check first with your local bee inspector what is currently permitted in your area and follow official guidance on their correct application.

Icing sugar (or powdered sugar, as it tends to be called in the USA) appears to work by causing mites to lose their grip on bees and fall off. Therefore, it is best used with mesh floors, which allow the mites to fall clear through to become ant food. Applying icing sugar (starch-free powdered sugar, not caster sugar) to a conventional, framed hive is generally done from above, trickling the sugar down between the frames, or by removing each frame, turning it on its side and dusting or sieving it over the bees. For obvious reasons, neither of these methods can conveniently be used on a top bar colony.

I had some success using an old smoker to puff icing sugar between combs, but the job really needs a proper tool. So I removed the bellows from the smoker and added a battery-powered blower, which blasts clouds of sugar upwards from below the screened floor at the push of a button, allowing me to treat my bees in seconds without even opening the hive. They show remarkably little reaction to this process and I don't think it does them any harm at all.

When you introduce a fresh swarm to an empty hive, I recommend that you give the bees a good dusting with powdered sugar. This will go a long way to removing any phoretic[50] *Varroa* that have hitched a ride with the swarm and helps the bees to get off to a good start before they acquire any more - as they inevitably will - through contact with flowers that other bees have visited, and from drones, which have a habit of wandering from hive to hive. This is a particularly good time to perform the dusting, as it is (hopefully) the only time you will have all the bees in that colony in a heap on the floor of your hive, with a good chance of every bees getting a little sugar powder on them. As the bees lick the sugar off each other, one can suppose that they may also take a nip at any lingering mites and just possibly this habit may continue once the sugar has all gone. The sugar itself will not be unwelcome as a small boost to their food supply before they start foraging in earnest. Make sure that it does not contain starch, which is injurious to bees.

The other occasion when a general sugar powdering is handy - beyond its use as a miticide - is when two unrelated colonies have to be combined, such as when one of them looks to be too weak to survive the winter, or is hopelessly queen-less. In this case, the bees become so occupied with removing the sugar that they forget to fight – or perhaps the powder disguises the differences in scent between the two colonies.

In the normal course of events, it is a good plan to blow or sprinkle powdered sugar onto your bees once per month in their first season and give them a good dusting before they settle down for the winter. It is preferable to apply the sugar when the maximum number of bees are at home, so dusk is a good time to do it. If you use my recommended method of blowing sugar from underneath, through the mesh floor, you do not need to disturb the brood nest at all. Users of vertical hives will have to use some ingenuity (possibly including a mechanical lift or an assistant) to perform this treatment, if they deem it necessary.

I must be clear here that *nothing* should be sprayed or otherwise introduced into the hive at a time when there is a possibility of it getting into the honey. Even powdered sugar should be managed carefully, as sprinkling it onto comb while nectar cells are open will lead to contamination of honey - albeit at a low level and with sucrose rather than, say, thymol, which would affect the taste and potentially cause toxicity issues.

This is a tricky issue, as I prefer to leave honey on over winter and harvest in spring, when fresh supplies are coming in. This means that, if I feed syrup or fondant (as I have had to do after the last two 'summers'), and the bees store it alongside their honey, I have a problem to separate the two. Ivy honey, for example, looks a lot like finely granulated sugar in cells. I have had combs with honey one side and sugar the other.

---

50 *Phoretic* mites are those on the bees themselves, rather than in cells, where they reproduce.

So we have to establish a protocol for any treatments we carry out, and any harvesting we undertake, such that the possibility of any contamination - even with relatively benign substances - is minimized and preferably eliminated.

This means that sugar sprinkling must be confined to periods when there is no open nectar or honey, and any lipophilic or volatile compounds, such as essential oils, must only be used when there is a reasonable expectation of the food they are added to being used by the bees fairly quickly (such as during a period of dearth) and not being stored alongside any honey we may intend to harvest.

This regime may impose some serious restrictions on us, and it may be that in some places and some seasons, honey for our own use or for sale will have to be taken as soon as it is capped, and winter stores made up with syrup.

SHAKEN SWARM

So-called 'shaken swarm' is a mechanical method by which the mite load can be drastically reduced in a colony which has come through its first winter in sound condition but with more mites than we – or the bees – would like. The whole colony is simply shaken from its over-wintered comb into a newly-prepared, empty hive, and left to start again from scratch. All brood comb is discarded, along with all the mites that were busy reproducing in the cells. Feeding is usually required, unless there is a strong flow on, which, I would say, is the best time to attempt this rather brutal manoeuvre. If it is timed to coincide with preparations for swarming, then 'shaken swarm' can be an effective swarm control measure, and if all goes well, the bees will throw themselves enthusiastically into the creation of a new home, just as if they had swarmed. Occasionally, however, they will take offence at being so roughly handled and abscond. There is an element of risk in all attempts to persuade bees that we know better than they do.

The presence of mites in such numbers as to cause concern can be detected by means of a piece of white plastic or card placed on top of the floorboard or mesh inside the hive. Coating it with a smear of grease made up from one part beeswax to five parts vegetable oil helps mites and other debris to stick to it and thus allows for easy counting.

In the context of parasites and diseases, you may like to consider the following words, written by A Gilman, an original thinker and author of one of my favourite beekeeping books, *Practical Bee Breeding*, published 1928.

"...disease is an expression of lowered vitality ...and simultaneously

with increased fecundity there has been an extraordinary increase of disease. Their connection may be denied, but when we find a similar occurrence taking place with other livestock which we know to have been pushed for super-production, we consider the matter far more than a mere coincidence."

"...the increase of diseases has occurred principally in those countries where modern methods of breeding have prevailed. In America, brood diseases became so devastating as to call for legislation... on the continent of Europe, apiarists have been troubled with Nosema disease... we had Isle of Wight disease, which so decimated apiaries all over the country that we had to resort to foreign bees for re-stocking purposes."

"...the only conclusion to which one can come, is that the principles on which the whole structure of modern apiculture are based must be at fault, in either one or more important directions."

He goes on to quote from Dadant's *System of Beekeeping* (pub. 1920) as follows:

"If anyone had asked us, twenty years ago, how much trouble might be expected from bee-diseases, we should probably have shrugged our shoulders and answered that they were very insignificant and hardly worthy of notice. For forty years after we began beekeeping the only disease we saw in the apiary was diarrhoea... from which the bees suffered more or less after a protracted winter, especially when their food was not of the best... Foul brood, in either of its two forms was entirely unknown (sic) to us. In 1903 the writer had to go as far away as Colorado to be able to see some rare samples of it... It was not until the spring of 1908 that we found it among our bees..."

So, Dadant himself never saw foul brood in his own bees until 1908 - just one hundred years ago as I write.

And yet, the beekeeping 'authorities' continued to preach the litany of 'movable frames and foundation' for another century! And what is more remarkable, is that they continue to do so despite declining bee health - and they still refuse to take seriously those who have turned their backs on that travesty of a beehive - the Langstroth - and are experimenting with protocols designed to help the honeybee return to its natural state, uncorrupted by synthetic medicines, ill-designed accommodation and ill-conceived breeding methods.

A Top Bar Hive Ready For A Swarm

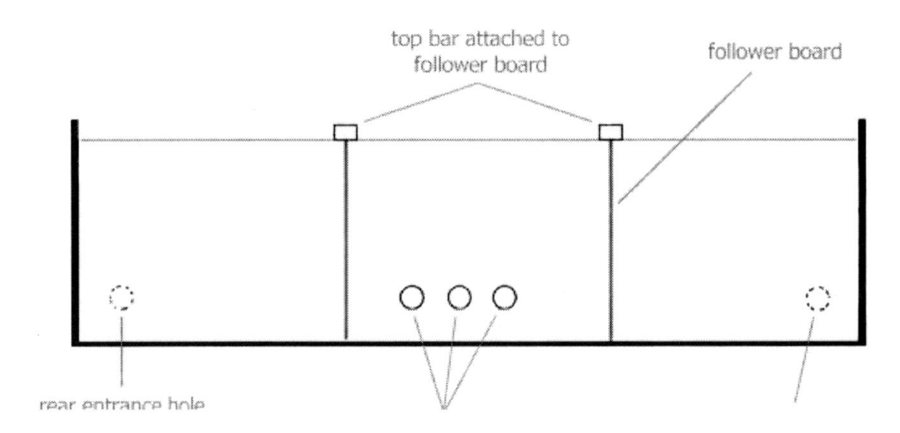

A Top Bar Hive With New Swarm

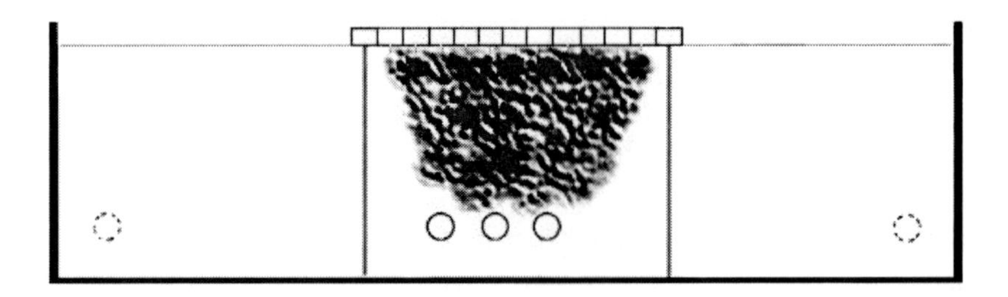

Expanding The Hive

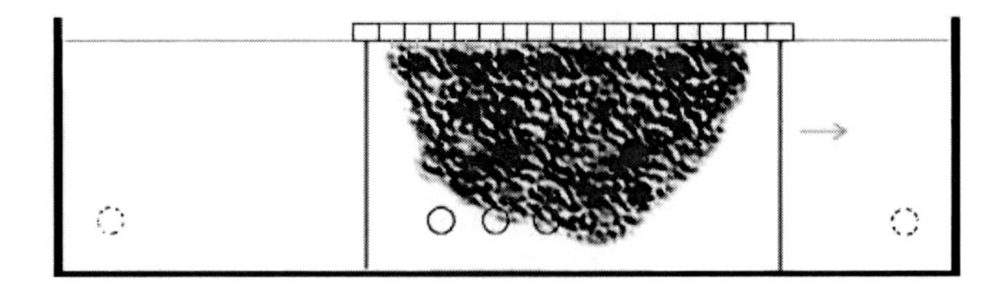

HARVESTING HONEY AND ESTIMATING WINTER STORES

One summer day, when you open your hive, you will find a sealed comb of honey just inside. This is where you have to start making judgements that will affect the bees' ability comfortably to over-winter.

Since you introduced them to your hive, they will have expended all their energy in building comb, feeding their young and providing themselves with stores to carry the colony through the winter. If the weather has been kind and there has been a plentiful supply of nectar, they will store more than they are likely to use. Depending on how the summer weather continues, which plants are flowering in your area and how warm the autumn turns out to be, they may or may not be able to replace any honey you take at this point and the length and severity of the following winter will dictate the actual amount they need to get by.

Commercial beekeepers - and most framed-hive amateurs - tend to take their honey all at once at the end of the season, by removing the supers from the hives and then feeding sugar syrup to compensate the bees for their loss. In places and times when there is an early-flowering, high-yielding crop such as oilseed rape (a.k.a. Canola), this honey may be taken earlier, with the possibility of a later, second crop, perhaps from the heather, if you have any nearby. A certain amount of honey will be stored in the brood chamber, which responsible beekeepers leave for the bees (commercial men often take this as well).

That is not how we do things with a hTBH (vertical TBHers do tend to remove honey by the box rather than by the comb).

We consider it as self-evident that if the bees store honey for their own use, then honey is what they want and need. Sugar syrup is, at best, a poor substitute. Therefore, we strive to leave enough honey in the hive for the bees' winter feed. However, as none of us can be certain what weather the winter will bring, we can only take our best guess and be prepared to help the bees out should they need supplementary feeding.

The actual process of harvesting honey is simple enough: take one sealed[51] comb at a time, cut it from the top bar and replace the bar on the outside of the storage area for the bees to build more comb on.

---

51 Bees seal or cap honey when they have reduced its water content to a level where fermentation cannot occur, i.e. less that 20%.

PROCESSING HONEY

To my taste, honey is perfect just as it is – straight from the comb. Until the invention of the centrifugal extractor some time in the late 19th century, that is how everyone ate their honey. Now, comb honey may be regarded with suspicion by some and a luxury product by others. It certainly fetches a much higher price; just the other day I saw some comb from New Zealand, fancily packaged and labelled 'organic', selling at over £6 for 250g, at least three times the price of extracted honey. (Is it not absurd that honey and apples are flown halfway around the world, when – with a sane agricultural system - in the UK we could probably be self-sufficient in both?)[52]

Top bar hives are perfect for producing comb honey and nothing is required of you other than to cut it into conveniently-sized chunks with a sharp, serrated knife on a clean board[53] and place them into suitable containers. Odd-shaped and smaller pieces can be put into normal screw-topped jars, topped up with liquid honey and sold as 'chunk honey' at a premium price.

If you or your customers (or family) demand at least some honey in liquid form, there are several ways you can extract it. The simplest is to toss comb into a stainless steel bucket, thoroughly mush it up with a paddle of some sort, then strain it through muslin. Make sure to do this where bees cannot get at it, or they will put it straight back in their combs.

Another method is to build or buy some sort of press. The type sold for small-scale fruit pressing will do a reasonable job, but you need a great deal of pressure to squeeze every drop of honey from the comb, as Brother Adam found when, during the 1920s, he tried to use a converted cider press to crush combs full of heather honey, which is thixotropic (jelly-like) and difficult to centrifuge. He realized that he needed something much more powerful and had built a monstrous hydraulic press, which applied a pressure of over 200 lb per square inch (31kg/cm^2), which worked well for more than 50 years, until 'Health and Safety' regulations finally got the better of it.

If you have an inventive mind, you may be able to construct a press

---

52 'I get very pissed off when people say "Oh, it'll be just like the war". In those days we had loads of orchards, we had loads of farmers, we did grow loads of stuff. At the moment, 70 percent of the wheat we grow goes to feed cattle which then feed us. And we have no fruit - the fact that Tesco and Sainsbury stock apples from New Zealand in September is a scandal, and that has to stop.'

Rosie Boycott, journalist, smallholder and Chair of London Food board, The Ecologist, Feb 2009.

53 Food hygiene regulations may require stainless steel surfaces for commercial production

using a car jack and some heavy timber, or, more simply, a long lever operating a plunger of some sort.

However you do it, pressing honey out of comb is a messy, tiresome business and I suggest that your energy is better spent persuading your customers of the benefits of honey in the comb. Point out to them that it retains all its vitamins and enzymes and full flavour, all of which are depleted by the extraction process and heat treatment that virtually all commercial and most amateur-produced honey is subjected to.

An early Greek top bar hive (photo: John Phipps)

Untreated, unheated honey in the comb that has had no chemical treatments is a premium product, so you can charge accordingly and this will more than make up for any shortfall in production that you may experience in converting from framed hives. I think it is always better to have a smaller quantity of a higher-value product, which you can stake your reputation on, that to be selling the same mediocre stuff that everyone else sells.

PREPARING FOR WINTER

No sun - no moon! No morn - no noon -
No dawn - no dusk - no proper time of day.
No warmth, no cheerfulness, no healthful ease,
No comfortable feel in any member -
No shade, no shine, no butterflies, no bees,
No fruits, no flowers, no leaves, no birds,
November!

Thomas Hood (1799-1845), in the poem called No!

The further they travelled north from the Equator, the more the honeybee had to adapt to the changing seasons. There are regions where the summer drought is much more of a nuisance to them than winter, while, at the northern-most limits of their territory, Siberian bees have only a few months in which to bring in stores that must last more than half the year.

An English beekeeper in southern Spain describes his local 'winter' conditions thus:

> *Right now it is nearly 3pm 18th January and the temperature outside is 22° C and full sun. The bees are out on the lavender and they keep brood rearing going through our winter. It can get cold at night but very rarely freezes. Of course the bees will cluster around the brood at night but fly freely during the day. There is plenty of forage right now, the almonds are in bloom and will continue into next month. February starts our swarming season here through till April. We don't need to prepare for winter here as there is enough forage but we do have to prepare for the summer dearth. Mid June through till mid September the heat dries up all the forage and we have to ensure our bees have sufficient stores to last them through this period. Water availability is paramount at this period also, they bring it in to evaporate and cool the hive.*[54]

Their enemy in the northern winter is not so much cold – bees can withstand quite severe drops in temperature[55] – as damp. Condensation inside a hive can wreak havoc with winter stores and turn everything mouldy, causing the bees endless problems and killing off many otherwise healthy colonies. I have seen this happen with commercial hives, but never yet with my top bar hives.

Conventional hives tend to be made of thinner timber than that which I

---

54 From the biobees forum, posted by 'Norm'

55 During an Advanced Beekeeping course at the National Science Laboratory, I saw bees that had been subjected to -75° C for an hour come back to life in the warm lab.

recommend for TBHs, which makes them vulnerable to condensation, as the wood does not provide enough insulation from outside temperatures, so that warm, moisture-laden air inside the hive condenses on the cool, inside walls. Also, having a shape and size that makes them difficult for bees to heat only exacerbates this problem.

The TBH design I recommend has several features that make it especially comfortable for bees in winter:

- Insulation can be added under the roof, directly on top of the top bars.
- Because the overwintering bees are contained in the middle section of the hive, they have follower boards and insulating air space either side of them, to which can be added solid insulation, such as straw or scrunched up newspaper, if required.
- The trapezoidal shape of the hive minimizes the volume of air they have to heat, compared to a vertical-sided, rectangular shape.
- Ventilation is provided through the floor, which in mild areas can be left open, although there is a good case to be made for a solid floorboard beneath the Varroa screen left permanently in place,so long as a small amount of ventilation is provided to prevent any accumulation of moisture.

Insulation, therefore, is not so much about keeping bees themselves warm, as providing the means to keep the interior surfaces of the hive warm so as to reduce condensation to a minimum, which makes the bees' job of keeping themselves warm so much easier.

WINTER STORES

Conventional wisdom has it that an average colony in a framed hive in Britain will need around 35lb of stores to get them through the winter. Estimates vary by at least 5lb either side and will in any case depend on your latitude and the variety of bees you keep. Climate change means that British winters are warmer overall than they were 50 or so years ago when these estimates were current, which, you may think, would mean that the bees need less food. In fact, bees consume *more* food in a mild winter than in a severe one, because in higher temperatures they remain active, while during a very cold spell, they go into what is more or less a state of hibernation and consume almost nothing.

Another variable is that some strains of honeybee are more 'thrifty' than others; a fact established by Brother Adam, who considered this trait to be most desirable, especially for the commercial beekeepers who measure feed in tonnes rather than kilos. He considered the British black bee to be one of the most able to spin out its stores through a long winter, while Italian bees rather less so, coming as they do from a region with shorter winters than ours.

This all means that we need to allow more food rather than less for a mild winter – except, of course, that if the winter is short, they will gather nectar later if it is available (I have seen ivy in flower with bees working it in mid-December) and start earlier on the pollen (in 2009, hazel catkins could be seen here in mid January, despite the unusually cold winter).

The answer to this conundrum may be to take only modest quantities of honey in summer, leaving a comfortable surplus for winter and then harvesting whatever the bees appear not to need in the spring, once you are sure that winter is really over. This is the protocol I currently employ.

Because the top bar hive is easy and relatively safe to open even in cool weather, it is a simple matter to check on stores on a fine day in winter, so long as you just peep into one end – the brood nest should on no account be disturbed in winter.

Drones (male bees). Note the rounded – and stingless - rear ends and large eyes. (photo: John Phipps)

SPRING AND SUMMER FEEDING

Under normal conditions, bees are more than capable of feeding themselves and only need help during unseasonal weather, such as a cold, wet spring, a hot, dry summer or when we interfere with their natural processes, such as by making splits or artificial swarming.

Some beekeepers practice 'stimulative feeding' in spring, supposedly to 'get the bees working'. To my mind, bees that need any form of persuasion to get busy in the spring are not worth keeping. In fact, I doubt that such bees exist in the wild, as laziness is not a quality favoured by natural selection. The real reason that bees are fed in spring is to encourage the queen to start laying earlier than she is naturally inclined to do, so as to have as strong a workforce as possible when the nectar flow begins for real. If the flow fails to arrive at the expected time, however, the bees will need more feeding just to keep them alive until it does.

Winter feeding should not be necessary if sufficient honey is left in the hive, but in poor seasons and long, wet winters, additional food may be needed. Which brings us to the Sugar Problem.

Ever since refined sugar became widely and cheaply available, beekeepers have regarded it as a fair substitute for honey for feeding to bees. Sugar syrup has been routinely poured into hives for so long that few question the wisdom of its use or consider that it could cause any difficulties for the bees.

A container placed within the hive can be used to feed bees if they run short towards the end of winter. Note the straw scattered on the surface to prevent drowning.

Many beekeepers are now beginning to question this assumption and to wonder if routine sugar feeding may be contributing to the overall poor state of bee health that is currently being reported.

I am not a chemist, so my comments on this matter are merely opinions, but the fact remains that refined cane sugar is not chemically identical to nectar or honey and that alone should make us wary of feeding it to our bees.

One way around this is, of course, to feed honey exclusively.

However, this has its own problems:

- If you feed liquid honey, you risk attracting the attention of other colonies in the area: bees can smell honey at a great distance and may descend *en masse* to rob out an already weakened colony.
- If you feed comb honey (in winter), you may have to disturb the nest to get it close enough to the clustering bees, which could cause a sudden and potentially fatal temperature drop as the colony breaks up and becomes defensive.
- If you feed honey from another source, you cannot be sure that it does not contain disease spores or traces of any chemicals that may have been used. Unless you are sure that it has not been heated, you also run the risk of it containing toxic (to bees) levels of HMF[56].

My solution is to do my best to ensure that the bees have sufficient stores at all times and to top up with syrup or fondant (a.k.a. 'candy': see below) only when deemed necessary. When I do so, I always mix in some of their own honey to make it smell and taste of something familiar.

STATION FEEDING

Given that all top feeders in the manufacturers' catalogues are designed to be used with framed hives, we as TBHers have a problem if it should become necessary to feed sugar syrup to our bees. One solution is to feed outside the hive, which is known as 'station feeding'. This consists of placing a suitable feeder where bees can fly to it and help themselves.

It is a convenient way to feed a number of colonies simultaneously, as there is no need to open hives, but the downside is that you may find yourself feeding bees from other apiaries in the area as well as your own. Some say that it also makes bees bad tempered and can

56 Hydroxymethylfurfural – toxic to bees – increases if honey is heated or aged. Some sources state that HMF may also be toxic to humans.

predispose them to robbing, but I can't say that I have noticed this.

A chicken drinking trough, such as the one illustrated, makes a good station feeder for thin syrup (1 kilo of sugar to 1 litre of water) with some straw floating in the surface to save bees from drowning.

A chicken waterer can be pressed into service as a station feeder for bees. Float fresh straw on the surface, so bees do not drown.

FEEDING WITH FONDANT

Fondant is made by boiling water and adding granulated, white cane sugar[57]. The recipe is 5lb of sugar to 1 pint of water (or 2kg to 500ml), in which I include about three tablespoons of organic cider vinegar[58]. Bring to the boil (should be around 234°F (112°C), or 'soft ball'), stirring constantly to avoid burning and when all the sugar is dissolved and the liquid is thick, syrupy and clear, add a good dollop[59] of honey and pour into a deep dish and allow to cool.

There are a couple of advantages in using fondant over sugar syrup,

---

57 Health foodies please note: you will kill your bees if you give them 'soft brown' or 'Demerara' or date syrup or anything else: they all contain stuff that bees cannot deal with! By all means, use organic granulated white sugar if you can afford it.

58 This has the effect of making the fondant slightly acid, which conforms more closely to the pH of nectar than sugar alone, which is neutral. Also, I suspect that cider vinegar may have other benefits, but cannot produce any supporting evidence for this yet.

especially in winter:

- It is semi-solid, which makes it easier to handle.
- You don't need special feeders to administer it.
- It does not stimulate queens into a premature laying frenzy, which syrup fed in late winter/early spring is apt to do.

In cool weather, when the bees are clustered together and unwilling to move far for food, should you find them with no honey nearby, you may be able to rescue them by pressing soft fondant into empty comb, close to the cluster. Otherwise, say during a period of famine caused by unseasonal weather, wrap lumps of fondant in cling film for transport, open them and place on the hive floor just inside the follower boards.

One thing you should never do is to feed your bees on imported honey or even honey from someone else's bees. If you do, you risk spreading diseases and even traces of other beekeepers' medications. Almost all liquid honey has been heat-treated, which can potentially make it toxic to bees (as well as destroying much of its nutritional value for us!)

If you feed honey, feed your own – preferably from the same hive.

I am experimenting – as are others - with various feeding methods, including using follower boards with built-in feeders. We will report all such developments and refinements on my web site at www.biobees.com.

---

59 This is not an approved metric measurement, but everyone knows one when they see it.

Bees building pure white comb on a top bar. At this stage, it is very delicate and is best not moved until it has had a chance to 'harden off'.

Bees collect pollen from a wide variety of flowers, given the opportunity. This colourful collection came from a hive in South Devon in June 2006.

YOUR SECOND BEEKEEPING SEASON

If you started beekeeping with one or two summer swarms, you may not, in your first year, have been able to harvest more than a taste of honey from your own hives. You will, I trust, have concentrated on gaining experience, reading books and observing bee behaviour.

As the anniversary of your first hiving approaches, you can consider some of the management procedures that you may want to attempt this season, such as artificial swarming and creating nuclei. You will also be hoping for your first real honey harvest.

BEE WATCHING

Now is the time to get used to watching the bees at the hive entrance and observing their behaviour, as many clues can be found about the health and status of a colony this way, without any disturbance to the bees.

One of the first things you will see in spring is pollen coming into the hive. Pollen is the bees' source of protein, as vital for raising their young as is the carbohydrate derived from nectar. If pollen is being carried in and the air is filled with a busy, contented hum and a steady stream of bees coming and going, you can be pretty sure that your colony has a laying queen and that all is well within.

During a nectar flow, you will be able to hear a constant hum from the hive at night, which is the sound of bees operating their own wing-powered air conditioning system to evaporate excess water from their stores. The water content must be reduced to less than 20%, or the honey will ferment and blow the caps off the cells. How the bees know to do this is one of those mysteries we may never solve: this is one of many instances of bees appearing to 'know' things that – given their brain size – they really have no right to know.

You may occasionally see an 'undertaker bee' fly out carrying the corpse of one of its sisters, which it will deposit outside. This is normal, hygienic behaviour and shows they are keeping the hive clean and tidy. If you see this happen more than once in a few minutes, you might want to take a peep in case there are more dead bees in the hive than is normal, possibly indicating a pesticide poisoning incident.

Sometimes you may see what look like mummified larvae being thrown out, which may well be an indication of near-starvation, especially if accompanied by a lack of foraging behaviour. Check for the presence of a queen and feed heavily if you see this behaviour.

Crawling bees on the ground beneath the hive may be a sign of virus infection, most likely resulting from a high mite population. This may be verified by removing the floorboard – if you have fitted a removable one – and examining the debris for fallen mites. If you have open mesh floors, you can assess mite drop by laying a sheet of paper over the floor, directly underneath the brood area. Don't leave it there for more than a day or so, or the bees will chew it up and throw it out the entrance.

Orderly, even-tempered behaviour is the hallmark of a healthy, fully functional colony. If what you see and hear is loud and chaotic, you may have a queen-less hive. Of course, this should never happen, but occasionally it does. Perhaps a virgin failed to return from her mating flight and the old queen had left or gone off-lay. Perhaps a supersedure[60] failed to produce a viable queen. Whatever the explanation, the colony needs re-queening as soon as possible and this is one time when you will bless the day you decided to make a couple of nucleus hives (see later). If you did not, then you will have to find a swarm cell from one of your other colonies, or put in a comb of freshly-laid eggs from another hive, or ring around your beekeeper friends for a spare queen. Introducing a new queen is best done by someone who has done it a few times before: beginners should seek help.

You should learn to spot drones. In spring, they are another indication of a healthy colony and may comprise between 5 and about 15% or a little more of the total force (I don't expect you to count them) and they look big-eyed and a little clumsy compared to the workers. They have no sting and are therefore ideal for practising your 'picking up' skills on. The presence of drones can be an early warning of swarm preparations: bees will not swarm without them.

Later in the season, from August onwards, you may see drones being evicted *en masse* from the entrance, or prevented by guards from entering. This is their fate: to be killed or forced from their home by their sisters before winter sets in and they become just another bunch of mouths to feed. If there are still drones present at the onset of frosty weather, this may be an indication either of queenlessness or imminent supersedure, which will probably fail, due to the unlikelihood of there being suitable conditions for mating. Unless you have a queen-right nucleus[61] handy, or someone nearby with the equipment and skill who

60 At any point in the life of a colony, the bees may decide that their queen must be deposed and replaced by a new one. Perhaps she is just getting on a bit and her egg-laying pattern has become poor. Perhaps she has been injured by a careless bee-keeper As bees have no retirement homes for ageing or ailing queens, they simply raise a new one and this is often accomplished without the bee-keeper even being aware of it, the only evidence after the event being an empty queen cell and a somewhat revived colony. When bees supersede their queen, they usually raise only one or two replacements, as opposed

can perform artificial insemination[62], the only action left to you is to combine that colony with another.

During the coldest part of the winter (if you have one), bees will remain house-bound and clustered, but on mild days they will perform cleansing flights to void their bowels. If the day is sunny but the temperature is low, some of them may not make it back home. This can be a particular problem with conventional hives that typically have entrances that allow quite a lot of light to enter, fooling bees into making hazardous winter flights. This is one of the many reasons that I choose to locate my entrances low down on the sloping sides, as in this position, direct sunlight cannot penetrate the body of the hive.

SPRING BUILD-UP

Once winter is over and done with and spring fills the air with the scent of fresh blossoms and all is well with the bees' world, things in the hive start happening rather fast.

At the first sign of a serious amount of nectar coming in, the queen goes into full-power laying mode and puts eggs into waiting cells at a fabulous rate – perhaps two or even three thousand eggs per day. This means, of course, that soon bees will be hatching at this rate too and the colony undergoes rapid expansion.

As this happens, you need to be aware of it and keep an eye on your colonies' growth, making sure to give them enough room by adding empty top bars inside the followers. You should err on the side of giving them more room that they need, but don't overdo it, as too much space could result in some rather 'artistic' comb building!

The first and greatest challenge is to find a way to handle the bees' natural desire to go forth and multiply: the swarming season is upon us.

---

to the dozen or more that may be started when they have swarming in mind. It also seems to happen later in the season. It is said by many bee-keepers that supersedure queens are the best you can have, as they have been 'elected' by the bees and not imposed by the bee-keeper in a re-queening exercise. Swarm queens are often superseded before the end of the same season.

61 A nucleus colony, or 'nuc', is simply a small colony created by the beekeeper, using a spare swarm queen and some bees from one or more larger colonies. It is useful to have one or more around as a

THE SWARMING IMPULSE

If there is one subject that occupies beekeepers' thoughts, dreams and conversations through late spring and early summer, other than the expected or actual honey flow, it is swarming and how to deal with it.

Whole books have been written about swarming and its control and/or prevention, yet, despite all the words that have been assembled and spoken, the inescapable fact remains that *bees will swarm as and when they choose to* – unless the beekeeper can somehow physically prevent them doing so, or persuade them that they have already swarmed by means of the various techniques of 'artificial swarming'.

2006 - A prime swarm, conveniently positioned - for once - for easy removal.

---

source of emergency queens.

62 AI is, in my opinion, one of the less welcome developments in bee breeding. It may have its uses in rescuing a strain threatened by extinction, but it generally produces poorly-mated queens that are rapidly superseded.

What is the problem with bees swarming?

Well, there is no problem for them, other than to find a new home for about half their number. This is what bees in the wild do: it is an essential part of their life-cycle to reproduce the colony by raising a new queen and for the old queen to vacate her palace along with about half of her children. The new queen, once she has mated, will inherit the old colony and proceed to raise her own progeny.

The most obvious drawback to this behaviour from the beekeeper's point of view is that you lose half your bees, just at the point in the season when they are most needed to make maximum use of the available forage. In short, if you just allow your bees to swarm willy-nilly, you risk having no surplus honey at all – and quite possibly, no bees, should secondary (cast) swarms follow the first (prime) swarm.

Swarm control is, some would say, the greatest challenge facing the beekeeper, unless his primary reason for keeping bees is to help replace the depleted feral colonies. I think there is a good case for allowing one or two colonies to swarm each year and take their chances in the wild: in terms of encouraging the long-term survival of honeybees, replenishing feral stocks is probably our best hope.

But for the most part, the best way of managing the swarming impulse from the point of view of practical beekeeping is to practise some form of 'artificial' swarming.

ARTIFICIAL SWARMING

This is simply a managed alternative to just letting a prime swarm emerge and then catching and hiving it, which is is often difficult and sometimes impossible, due to bees' annoying habit of swarming when you are not around and clustering way out of reach on the ends of branches in tall trees. Rather than risk losing your best colonies just before the honey season gets under way, it is advisable to take matters into your own hands, while allowing them to feel that they have fulfilled their natural swarming instinct.

Because even the smaller, 36" TBH is made to be more than twice the length of a standard 'National' brood chamber, you will usually have room to make an artificial swarm within the same hive as the mother colony, thus avoiding the need for additional woodwork and eliminating almost all lifting and carrying. This is particularly convenient if you intend to re-combine the swarm with the mother colony at a later date.

The procedure begins at the point when, during a routine spring inspection, you find one or more queen cells, sealed or about to be sealed, on the edges of the combs: a sure sign that swarm preparations are under way.

In fact, the first sign that a colony is thinking about swarming is when they begin to raise drones, as bees will not swarm without them. Because it takes 23-24 days from egg to adult for a drone to emerge and only 15-16 days[63] for a queen, you will notice drones developing before you find swarm (queen) cells. However, swarm cells – often half a dozen or more – are the sure sign that swarming is imminent. You will find them mostly on the outer edges[64] of combs near the brood-raising area of the hive.

Most of the old books (and too many newer ones) recommend destroying these queen cells as the primary swarm prevention measure. Why this technique has been promoted for so long is a mystery, as the plain fact is that *it does not work*. At best, it is a temporary measure that can buy you time in an emergency, but – as any experienced beekeeper will tell you – if a colony has decided to swarm, then they *will* swarm at the first opportunity and if all you do is to remove queen cells, they will become demoralised and fail to thrive.

The purpose of artificial swarming is to give them the feeling of having swarmed, but on your timetable rather than theirs.

So, when you see those tell-tale queen cells, regard them as a welcome opportunity to increase your stock with almost no effort.

Here is one way you can do the job, without using any extra equipment.

HIVE ROTATION METHOD FOR ARTIFICIAL SWARMING

I am assuming you have a strong colony in a TBH built to my design, i.e. with follower boards, three entrance holes near the centre and one each end on the opposite side. The colony, in early summer, may have eight or more combs of brood, two or three of which have swarm cells, either sealed or about to be sealed.

Carry out the following procedure on a sunny day, between midday and mid afternoon, when the flying bees are out foraging.

1. *Making sure that the laying queen is not on any of them*, lift out each of the combs with queen cells and place them, in the same

---

63 You will find the figures 24 and 16 respectively in most bee-keeping books. Bees raised on natural cells tend to emerge a day or so earlier.

64 Recent research suggests that queen cells benefit from being slightly cooler than the brood nest, which would be why bees construct them on the edges of combs rather than the centre and could go some way to explaining why queens raised artificially are often inferior to swarm queens.

order - *with adhering bees* - on the other side of the follower board, towards the end of the hive that has the most free space. If there are more than three combs with queen cells on them, *take only three* and remove the queen cells from the others, or use them to create nuclei or to re-queen a queen-less colony, if you happen to have one.

2. Transfer to the new colony you have just created, two combs with stores (nectar and pollen) and place *either side* of the others.
3. Close up the gaps in the main colony and place new top bars to the 'honey' side of the remaining brood frames so they can continue to develop their nest. Close the entrance down to one hole for up to 2 weeks, to help the reduced colony defend itself.
4. Add a couple of empty top bars to the new colony, then slide the whole set up against the end of the hive and open their entrance hole[65], which is on the opposite side of the hive to the main colony entrance. Add another follower board so you have an empty space between the two colonies.
5. *Now, with the help of a friend, rotate the hive through 180 degrees* so that the newly opened entrance faces the same direction that the main entrance previously faced.
6. The foraging bees will return to the new entrance as it is almost in the same position and looks identical to the old entrance, albeit only a single hole. Finding their sisters at home, they will continue to forage as if nothing had happened, perhaps a little puzzled by the changes, but not put off their duties. They are returning to a hive that is now queen-less, but with a new queen on the way.
7. Within 8 days (assuming at least one of the cells was sealed), a new queen will emerge in this new colony and kill off her rivals. If all goes well, a few days later she will go on her mating flight and return as the new mother of the colony. It is highly unlikely that she will lead out a swarm as she will have plenty of space for expansion: your job is to ensure that the workers have bars on which to build new comb, or move them into a new hive either very close by (within 3 feet) or in an apiary more than 2 miles away.
8. Meanwhile, the old queen in the original colony will continue to lay, or if she had already reduced her output in preparation for swarming, she will resume laying, so the original colony will continue to develop virtually without interruption: most of the

---

65 See hive building plans

guard bees will quickly become foragers. It will, if favourable conditions prevail, make a surplus by the end of summer. It is likely that the old queen will be superseded before winter, which is as it should be and allows the bees to decide on the timetable, which is their prerogative.

9. Bees born in the original colony after the rotation will, when mature, fly from and back to their new entrance, replacing the older flying workers who have returned to the new colony. The reduced population will, in almost every case, ensure that the old colony will not attempt to swarm again that season.

10. Once the new colony is well under way, the two groups of bees can, if desired, be re-combined simply by removing the follower boards and moving the end colony back towards the centre, harvesting any combs of honey in between. If the two queens should meet, they may fight to the death (the younger usually wins) or they may live quite peaceably together, laying side-by-side for some time. This happens quite routinely, even in framed hives.

You may want to re-combine the two colonies because one of them looks too small to over-winter successfully, or to take advantage of a strong flow later in the season. One really strong colony will always out-perform two weaker ones.

Note that, in the above procedure, everything happens as it would if the bees had swarmed themselves, except that the old queen stays where she is and the new queen hatches after the colony has been split. It would be possible to modify the process to more closely mimic nature, by moving the old queen and leaving the swarm cells in place, but then the foraging bees, finding their old queen still in place, may just decide to build more swarm cells and disappear anyway.

Bees, being inclined to vary their behaviour according to rules to which we are not privy, may decide to modify the above method of artificial swarming. A newly emerged queen may sometimes decide that swarming is quite a nice idea after all and disappear with some of your bees, despite your best endeavours.

Dividing The Colony

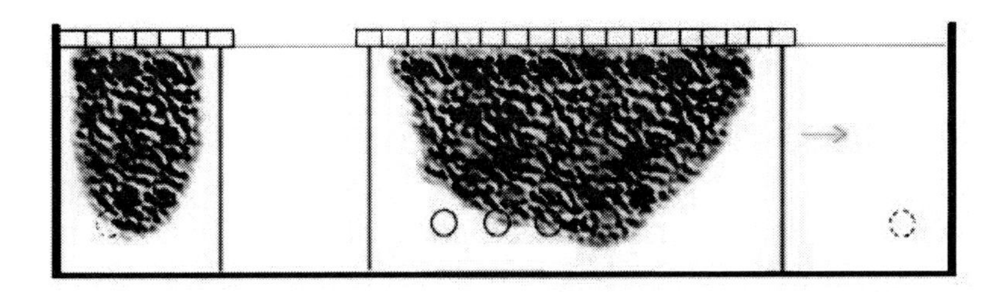

TWO APIARY METHOD OF SWARM CONTROL

If you have two apiaries, at least two miles apart, you can make an artificial swarm by removing several combs of brood with swarm cells and attached bees to the other location by means of a nucleus box, which is simply a short top bar hive – usually 5-6 bars long. By moving them at least two miles away, you reduce the risk of flying bees returning to their old home, especially if there is good forage to be had in between the two locations.

OTHER WAYS TO MANAGE SWARMING

Aside from artificial swarming, there are other possibilities for 'managing' the swarming impulse:

1. Make increase by dividing a colony before they show signs of swarming – ideal if you want to start more hives. Remove combs as for artificial swarming (except that they will not have queen cells, but must have some fresh eggs) and remove the newly-formed colony to a new hive with its entrance facing in a different direction from the original. Flying bees will return to the old hive, while the home bees will raise a new queen from an egg. Unless your queens are exceptionally productive and the conditions near to perfect, you will get little or no surplus honey from them in the same year.
2. Catch a newly-emerged swarm – risky and by no means guaranteed – but leaving empty hives in the area (primed with wax and propolis smells (and maybe a drop or two of lemon grass essential oil) will sometimes catch swarms without effort on your part. They may, however, decide that they prefer another neighbourhood and that is the last you will see of them.

Increasing your stocks by dividing strong colonies is a simple way to have more bees, as long as you do not expect to harvest honey from them until the following season.

You can do this by creating 'nuclei', which are nothing more than small colonies, headed by a newly-emerged queen, which will, given good foraging conditions, build up to full strength before winter. This is one operation that is ideally suited to my TBH – especially the longer variant – as it can be done in the same box using the extra entrances.

You can create a 'nuc' at either end of an already-occupied hive, or you can build a specialized hive with four entrances, placed on alternating sides. For a 3-nuc hive, you will need four dividers: two to hold a colony against each end and a further two to hold one in the middle.

You can also use the 'flowerpot hive' as a nucleus box, as its profile is similar to that of the TBH described in the supplement to this book ('How To Build A Top Bar Hive', available free from www.biobees.com).

Beekeeping is a process, not a destination. If you stay awake and listen to the bees, they will tell you what they need from you. Your job is to stay flexible: don't get bogged down in someone else's philosophy or techniques and always expect the unexpected. The bees will not disappoint you.

European bee houses use rear-opening hives. (photo: John Phipps)

TOP BAR STAND

This device is easy to build and a very useful 'third hand' for examining comb, especially when you need to do more than just look at it. I dreamed it up in bed one Sunday morning and built it before lunch out of odd bits and pieces lying around in my workshop. No doubt you can think of other ways of building such a device.

The top bar stand in action.

The length of the base is the same as that of a top bar – 17” - and 6” or so wide. The wire is about the same gauge used for coat hangers, bent to accommodate the width of a top bar. A slight inward 'spring' is an advantage, as this grips the comb and helps to keep the top bar in position.

Here it is on the bench...

...and folded for transport

MORE TOP BAR BEEHIVE IDEAS

Photos above are from a Polish beekeeper's site and reproduced by permission. http://homepage.interaccess.com/~netpol/POLISH/Ule/Wojtekshives2.html

As can be seen from the photo bottom left of the above group, it is easy to incorporate a window in one side of the hive, which enables you to see what the bees are doing without disturbing them. You will need to make a cover for the window so the bees don't overheat in summer or have condensation problems in winter. Similarly, you could use transparent follower boards.

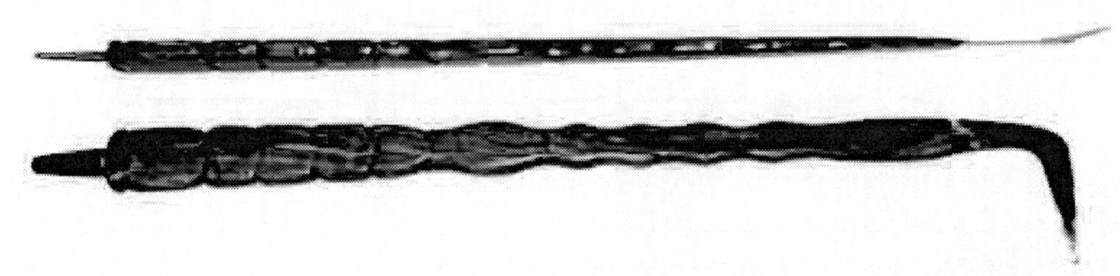

A long-handled knife, used to cut any attachments to side walls. (from the same source)

REFINEMENTS AND ADDITIONS

If you are of an inventive turn of mind, you may already have thought of some improvements and extra bits for your hive. Don't let me stop you, but do bear in mind that the essence of 'barefoot beekeeping' is simplicity: *resist the temptation to over-complicate.*

You might want to consider adding a 'landing board' for the bees, to make it easier to see what is going on around the entrance. I suspect this idea originated with Victorian beekeepers, who were fond of designing hives that resembled Georgian buildings (sometimes even having ornate columns either side of the entrance) but some people like them, even if the bees couldn't care one way or the other. A landing board – say 6” wide by 2” deep and 3/4” thick – could be added just below the central entrance holes, using a thin piece of wood as a support.

Pollen collectors are a useful means of gathering surplus pollen, either for one's own use or for drying and storing for the bees in case of dearth. In my area (south west England) there is rarely any shortage of pollen in the spring when bees need it most, thanks to plentiful willow, hazel and dandelions, but in other areas this may not be the case. I am still working on a design for a simple pollen collector and this will appear on my web site when ready, as will other notes and developments.

A 1” hole drilled through one of the follower boards and plugged with a cork provides an easy answer to feeding: a container with floating straw or a wooden float can be placed in the unoccupied end of the hive, the cork removed and the bees fed without any disturbance at all. It is even possible to feed honey this way without inviting robbing, due to the isolation of this area from the outside of the hive.

One of the most promising innovations is the use of top bars with cutaways to allow bees to move vertically into a super, to assist in maintaining straight comb building and to facilitate feeding. At first glance, this may seem to contradict the TBH principle of working at one level, and risking losing one of the big advantages of this hive: heat retention and exposing bees only at the ends of the colony. These objections can be overcome to a large extent by the use of 'condensation boxes' – an idea borrowed from the Warré hive – whereby shallow boxes with permeable mesh floors are filled with wool or sawdust are placed over the tops of the bars, covering the gaps. Roofs would, of course, need to be designed to accommodate the extra boxes.

Observation windows are worth the extra work, as a quick inspection can be made without disturbing any bees. Glass or Perspex/Plexiglass can be used; the window should be flush with the inside of the hive and a cover provided for insulation and to keep out light.

In writing this book, my intention has been not to create a prescriptive, rule-bound 'bible' of top bar beekeeping, but rather to suggest how it can be done, leaving space for you to think up your own solutions and experiment with your own ideas. I hope you now understand that beekeeping can be immensely absorbing, enjoyable, rewarding and at times frustrating, but it does not have to be either complicated or expensive.

Please keep an eye on www.biobees.com for updates and join our top bar forum to discuss all aspects of sustainable beekeeping.

Happy beekeeping!

from Phil's sketchbook

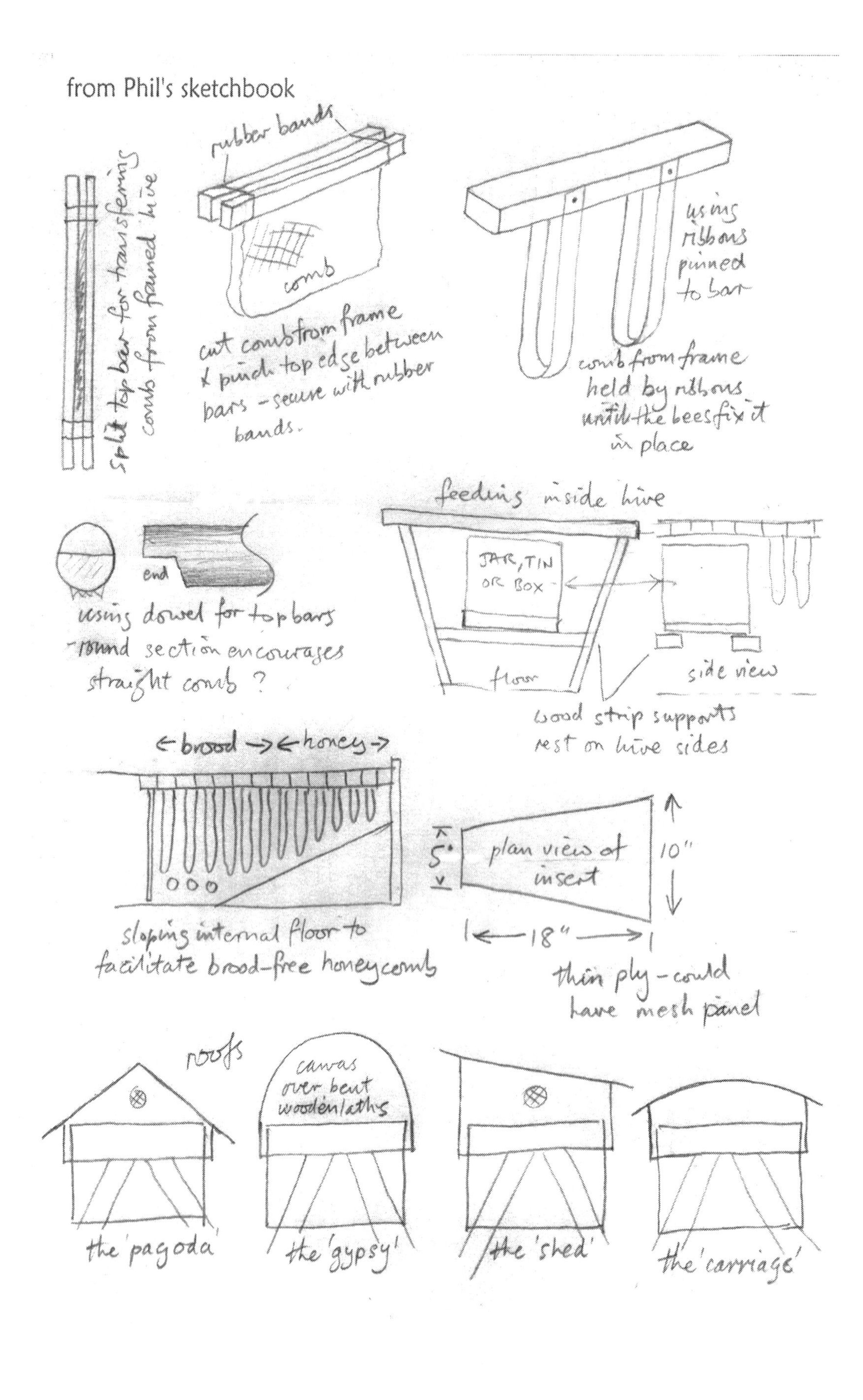

Alphabetical Index

(Due to the vagaries of the application I used in writing this book, some of the index entries above are somewhat eccentric. But then, so am I.)

AFTERWORD

In beekeeping, as elsewhere, there is no substitute for experience and this can only be acquired over time by careful and quiet observation and handling of bees. If you follow Bro. Adam's dictum, *"Listen to the bees"*, they will tell you what they need from you at any given time.

And always remember, the bees really could not care less about your welfare: their entire *modus operandi* is geared around ensuring the survival of their species, not ours.

Updates and future editions of this book will appear at

www.biobees.com

The author with a Warré hive

A NOTE ABOUT 'ORGANIC' BEEKEEPING

Many beekeepers reading this book, while happy to embrace the principles behind 'organic' farming and food, will find it difficult - if not impossible - to comply with the strict demands of Organic Certification.

Organic Honey is regulated by a strict set of guidelines, which covers not only the origin of bees, but also the siting of the apiaries. The standards indicate that the apiaries must be on land that is certified as organic and be such that within a radius of 4 miles from the apiary site, nectar and pollen sources consist essentially of organic crops or uncultivated areas.

Also enough distance must be maintained from non agricultural production sources that may lead to contamination, for example from urban centres, motorways, industrial areas, waste dumps, waste incinerators. The 4 miles guideline originates from research done by The National Pollen Research Institute, which is the maximum distance bee's travel from their hives.

These strict guidelines mean that is almost impossible for any UK producer to be certified as organic.

Source: http://www.beedata.com/news/organichoney.htm

Summary of the Key Standards for Organic Honey Production:

1. Siting of Apiary – must be on certified organic land that must not be treated with weed killers etc.

2. Hives – must be made of natural, untreated timber.

3. Conversion period – 12 months of organic management, during which time the wax must be changed to organic wax.

4. Origin of bees – 10% of the hives in an apiary can be replaced/increased using non-organic queens or swarms, provided that organic wax (from organic hives) is used. In this case the 12 month conversion period does not apply.

5. Foundation and comb – must be of organic wax, except when an apiary is first converted and organic wax is unavailable.

6. Foraging – for a radius of 3km (EU regs) or 4 miles (Soil Association) around the apiary, nectar and pollen sources must be essentially either

organic or wild/uncultivated. This area must not be subject to significant sources of pollution from e.g.: motorways, urban centres, incinerators, etc.. This is the only significant difference between the EU and the Soil Association standards in the area of honey production.

7. Feeding – must be organic honey or organic sugar and this may only take place between the last honey harvest and 15 days before the first nectar flow. (Crystal balls required, presumably.)

8. Disease control – similar to other livestock husbandry, the priority is to build up good health and vitality through positive management practices. Homoeopathic and herbal treatments and natural acids (Lactic, acetic, formic and oxalic) may be used without restriction. Other medication requires veterinary prescription, the wax must be replaced and there must be a withdrawal period of one year.

9. Queen rearing – artificial insemination is allowed but wing clipping is prohibited.

10. Extraction and bottling – no requirements beyond the normal measures to ensure separation and product integrity.

Source: http://www.beekeeping.org.uk/NL/info/info3.html

(The above are from UK sources; US, CA and other countries' standards may vary.)

The overall message is: IN ORDER TO SELL YOUR HONEY AS 'ORGANIC', you MUST be certified according to the prevailing Standards for Organic Certification.

You will also have to pay a not inconsiderable sum to the certifying body for the right to use the word 'organic' on your labels.

This last item effectively means than no small-scale honey producers will be able to sell their honey as 'organic', as (a) it will either be impossible to meet the standards economically, or (b) the fees for certification will be prohibitive.

So, how does the smallholder, who cannot comply with organic certification standards because of the costs or because of the geography of their area, legally market their produce so as to convey to their customers that it is 'naturally-grown' or 'chemical-free'?

This was the problem that long troubled smallholder Sky McCain, and he came up with a creative solution: the Wholesome Food Association. He quickly persuaded me that this was a great idea and we developed a set of principles and guidelines by which members would agree to grow and sell their produce.

The Wholesome Food Association (WFA) has members across the UK - mostly smallholders and small-scale farmers who grow quality produce without the use of synthetic chemicals - often imposing higher standards on themselves than would be required if they were 'certified organic'.

The WFA welcomes small-scale honey producers into their membership, and invites UK and EU beekeepers who wish to use the WFA label to sell their honey to join them.

See http://www.wholesomefood.org for more information.

Beekeepers in the USA have a similar organization: Certified Naturally Grown who have some specific standards for honey production - see http://www.naturallygrown.org/honeystandards.html

Here is an extract from a press release recently published by the WFA:

WHOLESOME HONEY BACK ON THE MENU

Raw, untreated honey, served in the comb, used to be the norm – and many believe that this is the way honey should be eaten.

Better still, honey that is guaranteed to come from bees that have never had any synthetic chemicals in their hive is once more available under the label of the Wholesome Food Association, which has been promoting locally produced, chemical-free food since 1999.

WFA Managing Director, Sky McCain says, "We want people to be able to buy locally-grown, wholesome food from people they trust to do the job well. Local, certified organic honey is virtually impossible to buy in the UK – it is almost all imported – so we are pleased that in some areas we can now offer a locally-made honey that has been produced to our chemical-free standards."

Raw, untreated honey is mostly produced by beekeepers who use 'top bar hives' – a low-tech, and often home-made hive that enables bees to build honeycomb to their own design, rather than to the pattern dictated by the pre-formed wax 'foundation' used in conventional hives.

Philip Chandler, author of 'The Barefoot Beekeeper', is pioneering this style of beekeeping in Britain. He comments, "Honeybees have been suffering for the last 150 years from the same sort of abuses as other factory-farmed animals. They have been badly housed, overworked, over-medicated and are now dying out as a result of this abuse and widespread poisoning of the land by pesticides. We want to sound the alarm now, before it is too late, and show how bees can be kept in a more natural way, without the need for chemicals to keep them alive."

"Beekeepers who follow chemical-free practices will welcome this initiative by the Wholesome Food Association and the public will, we hope, welcome the opportunity to be able to buy honey that is as pure as bees can make it."

Some pretty - if rather fanciful - Victorian hives

This illustration and the one on the next page are taken from an old book on entomology – I do not have the exact reference.

Bees 'daisy-chaining' as they measure up for new comb

VIEW OF THE INTERIOR OF A HUBER HIVE.
Four days after the introduction of a Swarm.

Read, every day, something no one else is reading. Think, every day, something no one else is thinking. Do, every day, something no one else would be silly enough to do. It is bad for the mind to be always part of unanimity.

Christopher Morley

And three people do it, three, can you imagine, three people walking in singin a bar of Alice's Restaurant and walking out. They may think it's an organization. And can you, can you imagine fifty people a day, I said fifty people a day walking in singin a bar of Alice's Restaurant and walking out. And friends they may thinks it's a movement.

Arlo Guthrie, Alice's Restaurant